SpringerBriefs in Ecology

SpringerBriefs present concise summaries of cutting-edge research and practical applications across a wide spectrum of fields. Featuring compact volumes of 50 to 125 pages, the series covers a range of content from professional to academic.

Typical topics might include:

- A timely report of state-of-the art analytical techniques
- A bridge between new research results, as published in journal articles, and a contextual literature review
- A snapshot of a hot or emerging topic
- An in-depth case study or clinical example
- A presentation of core concepts that students must understand in order to make independent contributions

More information about this series at http://www.springer.com/series/10157

Cang Hui • Pietro Landi
Henintsoa Onivola Minoarivelo
Andriamihaja Ramanantoanina

Ecological and Evolutionary Modelling

Cang Hui
Department of Mathematical Sciences
Stellenbosch University
Stellenbosch, Western Cape, South Africa

Pietro Landi
Department of Mathematical Sciences
Stellenbosch University
Stellenbosch, Western Cape, South Africa

Henintsoa Onivola Minoarivelo
Department of Mathematical Sciences
Stellenbosch University
Stellenbosch, Western Cape, South Africa

Andriamihaja Ramanantoanina
Department of Mathematical Sciences
Stellenbosch University
Stellenbosch, Western Cape, South Africa

ISSN 2192-4759   ISSN 2192-4767   (electronic)
SpringerBriefs in Ecology
ISBN 978-3-319-92149-5   ISBN 978-3-319-92150-1   (eBook)
https://doi.org/10.1007/978-3-319-92150-1

Library of Congress Control Number: 2018942940

Printed on acid-free paper

This Springer imprint is published by the registered company Springer International Publishing AG part
of Springer Nature.
The registered company address is: Gewerbestrasse 11, 6330 Cham, Switzerland

Preface

Ecology is the scientific study of the abundance and distribution of species, the interactions among organisms and feedback from their environments, and was known in previous decades as natural history. It concerns the fluctuation and entanglement of genes, populations, species and ecosystems, from a single pond to the entire planet. The philosophy of ecology runs deep in the veins of human civilisation, which has tried to identify humanity's position and role in nature, spanning many ancient philosophies that treated nature with awe and as an equal power, and religions that considered nature as something to be subdued, controlled and ruled over. Modern ecology blossomed in the eighteenth century, its champions Carl Linnaeus and Alexander von Humboldt, followed by Charles Darwin, then Arthur Tansley and Charles Elton, and reached a golden age in the era of Robert MacArthur, with many who remain active in the field.

Facing an unprecedented rate of biodiversity loss, we need to ponder upon the origin, function and adaptive changes of biodiversity – to date still the only known whole living system. Erwin Schrödinger asks, what is life? The functioning of adaptive, complex, living systems requires a grand framework to define it. All living things are made of cells, and fossil evidence suggests the first true cells were submarine hydrothermal microorganisms, which first inhabited the Earth 3.7 billion years ago. Once the seed was planted, nature then moulded the living form for numerous functions, branching into a grand tree of life. With the first fossil of comb jellies, Ctenophora, dating back 550 M years and the first brain structure in worms 500 M years ago, we are currently living together with more than 10 M strong species – not counting those lifeforms now extinct. How does such grandeur of biodiversity originate, interact and share limited habitats and niches on the cosmic pixie-dust of our blue planet? This is the mainstream of ecology: it is about the persistence and coexistence of species, the structure and function of biodiversity.

As the stewards of nature, contemporary ecologists are facing mounting challenges to ensure crucial ecosystem services through biodiversity monitoring and conservation, by sustaining interactions and feedback among multiple components and processes that foster species persistence and coexistence. Supported by Big Data from field surveys and laboratory experiments, from Earth observation systems and molecular biology advancement, and empowered by striking computational power, ecological modellers are now for the first time able to weave the threads of our knowledge with data into models across a grand range of scales and complexities, enabling true comprehension and valuable forecasts. This calls for many talented scholars to join forces and take on the global endeavour of ecological modelling and biodiversity informatics.

Mathematics is the most beautiful conjecture and the most powerful tool born of human ingenuity. It is without doubt that mathematical minds have worked throughout the early conceptualisation and later development of ecological theories. However, the walls between disciplines have hampered the flourish of this inter- and transdisciplinary field. Mathematicians often have vague concepts of ecology and evolution, while ecologists are becoming increasingly aware of the need for mathematical and quantitative methodologies in their research. As such, we introduce in this book a set of key concepts and modelling techniques in ecology and evolution. Topics covered in this book are by no means complete, but we hope they can serve as bridges to bring the possibilities of mathematics and ecology together. We hope that readers will consider not only the essential concepts in ecology and evolution but also the various standard mathematical and numerical tools used for exploration. To this end, the book can be considered an ensemble of selected topics to facilitate both ecological understanding and mathematical implementation. Due to the contextual dependence of ecological systems, we hope that like-minded scholars can learn the philosophy and procedure for model formulation, eventually to derive their own models for related systems. Let it be a hitchhiker's guide to both fields.

Throughout our writing journey in the past year, we have been greatly inspired by many colleagues to whom we own our appreciation (in alphabetical order): Åke Brännström, Richard Condit, Fabio Della Rossa, Fabio Dercole, Ulf Dieckmann, Gordon Fox, Klaus von Gadow, Laure Gallien, Alessandra Gragnani, Jessica Gurevitch, Fangliang He, Bill Kunin, Guillaume Latombe, Pierre Legendre, Zizhen Li, Jingjing Liang, Meloide McGeoch, Aziz Ouhinou, Anton Pauw, Carlo Piccardi, Dave Richardson, Sergio Rinaldi, James Rodger, Anna Traveset, James Vonesh, Feng Zhang, and all the members of the Mathematical Biosciences Hub at Stellenbosch University. We are also grateful to our many research grants and fellowships awarded to us by the National Research Foundation of South Africa, DST/NRF Centre of Excellence for Invasion Biology, DST/NRF Centre of Excellence for Mathematical and Statistical Sciences, Stellenbosch University subcommittee B, African Institute for Mathematical Sciences in Cape Town, National Science Foundation of China, Australian Research Council, and Deutscher Akademischer Austauschdienst (DAAD). We would also like to thank our families and friends for their support: CH thanks Beverley, Keira and Zachary; PL thanks Cecco, il Lando, Roland e Adele; HOM thanks Ando and Vahatra; AR thanks Andry

Kilome; all of us thank Vanessa du Plessis, Jonathan Downs, Sanjana Sundaram, Anthony Dunlap and Janet Slobodien for logistics. We hope you enjoy the book as much as we do.

March 2018 at the peak of the drought,   Cang Hui
Stellenbosch   Pietro Landi
                                                      Henintsoa Onivola Minoarivelo
                                                      Andriamihaja Ramanantoanina

Contents

Chapter 1
Biodiversity

Abstract Biodiversity is the most striking phenomenon in nature but perhaps also the most difficult to monitor and hypothesise. This chapter introduces key concepts and metrics for describing biodiversity patterns, as well as changes in these patterns. It starts with introducing the concepts of occupancy and aggregation across spatial scales for single species, followed by measures of species association and co-occurrence. It then discusses biodiversity patterns based on the manipulation of species-by-site matrices, from occupancy frequencies to species turnover and partitioning. It ends with the effects of imperfect detection and sampling on observed biodiversity patterns. This chapter lays the platform for understanding concepts and models of other chapters.

1.1 A Glimpse

Classification is the root of all scientific inference. Humans have been classifying species since antiquity. Aristotle described 500 species based on one of the oldest species classification systems. Carl Linnaeus formalised the hierarchical classification scheme of cellular organisms, including eight taxonomic levels, from top to bottom: domain, kingdom, phylum, class, order, family, genus and, finally, species. Each species is named using binomial nomenclature, also called the scientific or 'Latin name' (though they can include other languages, such as Greek). For instance, *Sturnus vulgaris* is the Latin name of European starlings, with *Sturnus* the genus name and *vulgaris* the specific epithet. Species are divided into six kingdoms: Bacteria, Fungi, Animalia, Protista, Plantae and Chromista. According to the World Conservation Union (2014), the most familiar kingdom is Animalia, including 67 K vertebrates and 1.3 M invertebrates. Plantae are largely groups of green plants with increasing levels of complexity, from numerous algal mats, 20 K bryophytes, 1.3 K lycopods and 10 K ferns to 1 K gymnosperms of conifers and cycads and nearly 300 K angiosperms of flowering plants. There are more than 1 M documented living species but nearly 10 M estimated undocumented, with some 5 billion species having dwelt on Earth throughout its history (Mora et al. 2011). Those living species

C. Hui et al., *Ecological and Evolutionary Modelling*, SpringerBriefs in Ecology, https://doi.org/10.1007/978-3-319-92150-1_1

represent 5×10^{37} copies and 50 billion tons of DNA base pairs (comparing to the 20 M words of all human language vocabularies). This is 1.25% of the 4 trillion tons of carbon of the entire biosphere. In pure weight, humans contribute 300 M tons – which is comparable to no more than the weight of all ant species. Amazingly, all organisms on Earth share at least one set of 355 genes (Weiss et al. 2016). Viruses are not included in this count; they are not genuine living organisms but replicator machineries of genetic material coated with proteins and sometimes also lipids.

Most species spread over only a limited geographical area, but there are a few which are truly ubiquitous. We can classify geographical regions into *biomes* based on the composition of their plants and animals, climates and habitats (e.g. tropical rainforests), where all organisms in one region are called a *biota*. An *ecotone* sets the boundary of two adjacent biomes. Anthropogenic biomes were proposed recently to reflect the level of human disturbance across the globe (e.g. dense settlements and croplands) (Ellis and Ramankutty 2008). The World Wildlife Fund has produced the Global 200 list of 867 ecoregions (Olson and Dinerstein 1998), labelling each with their conservation status: critical or endangered, vulnerable, relatively stable or intact. Based on high endemism and large portions of endangered species, a list of 33 biogeographical regions are given the status of biodiversity hotspots (e.g. the Cape Floristic Region in the southern tip of Africa) (Myers et al. 2000). The diversity of biological organisms and their interactions, with each other and with their environment, are essential for the provision of ecosystem services by which human societies are sustained. The Millennium Ecosystem Assessment (2005) has divided these services into four categories: for supporting (e.g. nutrient recycling), provisioning (e.g. food, water and medicine), regulating (e.g. climate regulation) and cultural (e.g. spiritual and recreation) functions. For instance, modern societies rely heavily on the energy resources of coal and oil, which are fossilised plants and animals deposited in the deep past. The rise of humanity has inevitably interrupted the natural flow of energies and materials in the biosphere and has led to the current so-called age of extinction, known as the Anthropocene. This chapter presents selected topics on structures and patterns of biodiversity and leaves models on the dynamics, genesis and functioning of biodiversity in other chapters.

1.2   Occupancy-Abundance Relationship

The simplest data in ecology is the point pattern of locations of individual organisms in a particular area (Fig. 1.1a). If we lay a grid (with the size of each grid cell *a*) over the point pattern, some grid cells will be left empty while others become occupied by a number of individuals (Fig. 1.1b). Similarly if we locate a small random circle of size *a* in the point pattern, the number of individuals occurring in the circle will follow a probability distribution. If the distribution of each organism is completely random and independent from any other organisms, we could use a Poisson process,

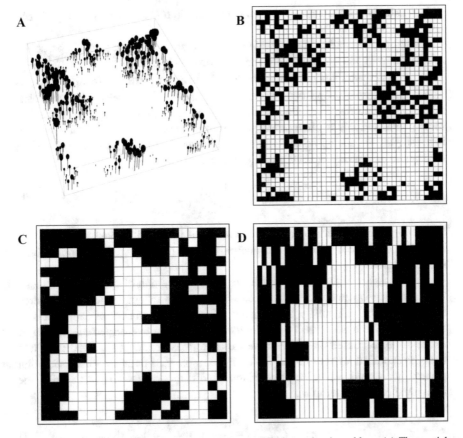

Fig. 1.1 An illustration of point patterns and the modifiable areal unit problem. (**a**) The spatial pattern of a population of plants in a box generated from the individual-based model (Hui et al. 2010, 2013). The presence-absence binary maps at the grain resolution of (**b**) 40 × 40, (**c**) 20 × 20 and (**d**) 40 × 10 grains (Redrawn from Hui (2009a))

the simplest point process model, to describe this point pattern. In this case, the probability of having n individuals in a sample can be described as follows:

$$p_a(n) = \frac{(d \cdot a)^n \exp(-d \cdot a)}{n!}, \tag{1.1}$$

where d is the population density. Note, this probability is both a function of population density and the sampling grain size a (signalling the effect of spatial scales). For such a point pattern, we could estimate the number of occupied grid cells (defined as the occupancy)

$$p_a^+ = 1 - p_a(0) = 1 - \exp(-d \cdot a). \tag{1.2}$$

This is called the occupancy-abundance relationship (OAR) for a randomly distributed species (Wright 1991).

Given the knowledge of how wide a species distributes in an area at a particular spatial scale (p_a^+), we could estimate the population density as $d = (-1/a) \cdot \ln(1 - p_a^+)$ and the total number of individuals in an area of size A as $N = d \cdot A$. This is useful as occupancy and occurrence can be obtained with relatively low costs, whereas the information of abundance requires counting literally all inhabiting individuals, and this is often unfeasible. Another application of this Poisson OAR is to estimate the demographic changes in the mean abundance ($\mu_a \equiv d \cdot a$) based on changes in occupancy or occurrence (Hui et al. 2012):

$$\Delta\mu_a = \frac{1}{1 - p_a^+}\Delta p_a^+. \tag{1.3}$$

It is evident that the demographic trend $\Delta\mu_a$ and the distributional trend Δp_a^+ for randomly distributed species are synchronised – expanding species implies abundance increases.

The suitability of the Poisson OAR lies in its assumption of species randomness, which is often violated in reality. Species distributions are aggregated and clustered in space, which can inflate the variance of observed numbers of individuals in samples σ^2. We could classify point patterns into being overdispersed, random or under-dispersed based on whether the variance σ^2 is greater than, equal to or less than the expected number of individuals in a sample μ_a. For an overdispersed point pattern, we could use a negative binomial distribution (NBD) to describe the individual counts in samples:

$$p_a(n) = C_{k+n-1}^n \left(\frac{\mu_a}{\mu_a + k}\right)^n \left(\frac{k}{k + \mu_a}\right)^k, \tag{1.4}$$

where C_m^n is the binomial coefficient of m choose n, $m! \, /(n! \, (m - n)!)$; k is the clumping parameter, ranging from highly aggregated ($k = 0$) to random ($k = +\infty$). As k tends to infinity, the NBD converges to a Poisson distribution. Similarly, we can have the following OAR for overdispersed species (He and Gaston 2000):

$$p_a^+ = 1 - \left(1 + \frac{\mu_a}{k}\right)^{-k}. \tag{1.5}$$

Note that the variance of counts for a Poisson process is $\sigma^2 = \mu_a$ and for an NBD process $\sigma^2 = \mu_a + \mu_a^2/k$. This is still different from the empirical variance-mean relationship known as Taylor's (1961) power law, $\sigma^2 = c \cdot \mu_a^b$.

A number of modifications have been proposed so that an OAR can emerge from a point pattern with its variance-mean relationship following Taylor's power law. For instance, an OAR considering Taylor's power law is (He and Gaston 2003):

$$p_a^+ = 1 - \left(\frac{\mu_a}{\sigma^2}\right)^{\frac{\mu_a^2}{\sigma^2 - \mu_a}}. \tag{1.6}$$

When the occupancy has changed, we could again derive the demography-distribution trend (Hui et al. 2012):

$$\Delta\mu_a = -\frac{\mu_a w}{1 - p_a^+}\frac{1}{\theta}\Delta p_a^+, \tag{1.7}$$

where $w = (\mu_a - \sigma^2)/\mu_a$ and $\theta = (b-1)\mu_a + [(b-1) - (b-2)w]\ln(1 - p_a^+)$. Clearly, to estimate the demographic trend for overdispersed species, knowing only the occupancy and distribution trend is not sufficient. Moreover, for overdispersed species ($w < 0$), if $1 < b < 2$, we have $\theta > 0$ and thus $\Delta\mu_a \sim +\Delta p_a^+$; if $0 < b < 1$ or $b > 2$, we have $\theta < 0$ and thus $\Delta\mu_a \sim -\Delta p_a^+$. Therefore, for most overdispersed distributions ($1 < b < 2$), the demographic trend has the same direction as the distribution trend, although with notable exceptions.

We could also formulate the OAR based on other mathematical frameworks. For instance, let there be N individuals over a landscape of size A that is divided into M number of grid cells with each size a. The probability that a cell contains n number of individuals can be derived from the maximum entropy theory. The entropy of a probability distribution is defined as

$$H \equiv -\sum_{n=0}^{N} p_a(n) \cdot \ln(p_a(n)). \tag{1.8}$$

Knowing N individuals in M cells is equivalent to assuming the constraints of $\sum_{n=0}^{N} p_a(n) = 1$ and $\sum_{n=0}^{N} n \cdot p_a(n) = N/M \equiv \mu_a$. The solution to this entropy maximisation under constraints is a geometric distribution (Park and Bera 2009):

$$p_a(n) = \frac{1-r}{1-r^{N+1}} \cdot r^n \approx (1-r)r^n, \tag{1.9}$$

where $r = \mu_a/(1 + \mu_a)$. This then leads to a distinct OAR, $p_a^+ = r$; note how the OAR also changes with sampling grain a. The geometric distribution is a special case of the discrete compound Poisson distribution. It is also a special case of the NBD, as the sum of many independent geometrically distributed variables converges to an NBD.

1.3   Aggregation Metrics

The generally non-random, aggregated spatial distribution of individuals has been of interest for long. Such non-randomness emerges from both the spatial heterogeneity of a habitat and the nonlinearity of biological processes, such as density-dependent growth and dispersal, as well as organism-environment feedbacks at fine scales. Synonymous terminology exists in ecology literature such as aggregated, overdispersed, clustered, clumped, autocorrelated, contagious and patchy, to name a few. As mentioned earlier, spatial variance (σ^2) becomes the first intuitive way to describe the spatial heterogeneity of species distributions. Based on the variance σ^2 and the mean μ_a of sampled abundance, a group of spatially implicit indices was proposed, including the variance-mean ratio (also called the coefficient of diffusion), Morisita's (1962) I_M, Lloyd's (1967) I_L, the clumping parameter k in the NBD and the exponent b of Taylor's power law. With the measures of species distributions shifting towards spatial structure, local indicators of spatial autocorrelation (LISA) statistics (Anselin 1995) have been proposed to capture the level of spatial autocorrelation, such as Ripley's (1976) K function, Moran's (1950) I, correlograms and join-count statistics (Cliff and Ord 1981). Spatially explicit indices are also available, such as Perry's (1995) SADIE and Peleg et al.'s (1989) Earth Mover's Distance (EMD). Besides these indices of spatial statistics, fractal dimensions, lacunarity and area of occupancy have also been used to describe the spatial structure of species distributions. A proper categorisation of these metrics was provided in Hui et al. (2010).

Let us now explore the meaning of spatially semi-explicit aggregation. As mentioned in Sect. 1.2, the simplest format of reporting a point pattern is the presence-absence map as often used in species atlases. This binary format map is mathematically equal to a binary matrix, the element being either 0 (absence) or 1 (presence). The simplest way of describing the spatial structure of a binary matrix is the join-count statistics, where the *global density* of a population is defined as the probability that a randomly chosen cell in this binary map contains individuals of the focal species. It has the same meaning as species occupancy, p_a^+. The *local density* of a species $q_a^{+|+}$ is the conditional probability that a randomly chosen cell adjacent to an occupied cell is also occupied, which is the simplest LISA metric. There are another three conditional probabilities regarding the state of a focal cell and its neighbouring cells: an occupied cell with an empty neighbour, $q_a^{0|+}$; an empty cell with an occupied neighbour, $q_a^{+|0}$; and an empty cell with an empty neighbour, $q_a^{0|0}$. In effect, we only need two variables, p_a^+ and $q_a^{+|+}$, to express all other probabilities. For instance, we have $q_a^{+|0} = \left(1 - q_a^{+|+}\right) p_a^+ / \left(1 - p_a^+\right)$. Another two inequality conditions constrain the feasible range of these two variables: $0 \le p_a^+ \le 1$ and $2 - 1/p_a^+ \le q_a^{+|+} \le 1$. Spatial clustering can be defined as $p_a^+ < q_a^{+|+}$, representing a positive local spatial autocorrelation, and the random distribution can be defined as $p_a^+ = q_a^{+|+}$.

1.4 Scaling Pattern of Occupancy

In the analysis of ecological patterns across scales (or called scaling patterns), a severe statistical problem emerges, i.e. how the measurements of those spatial patterns depend on the scales in question, formally named the modifiable areal unit problem (Openshaw 1984). The MAUP consists of two parts: the scale problem and the aggregation problem (also called the zoning problem) which, respectively, are caused by the changing of the size and the shape of grain (From Fig. 1.1b–d; Hui 2009a).

Let us consider species distribution when combining four adjacent square grid cells (with coordinates $\{x, y\}$, $\{x + 1, y\}$, $\{x, y + 1\}$ and $\{x + 1, y + 1\}$) into one larger cell, i.e. increasing the grain from a to $4a$. To calculate the probability of absence in a $4a$-size cell p_{4a}^0, we let the probability of absence p_a^0 for a randomly chosen subcell $\{x, y\}$ multiplied by twice the conditional probability of its adjacent cells ($\{x + 1, y\}$ and $\{x, y + 1\}$) also being empty $q_a^{0|0}$ and then multiply by the probability that a chosen subcell $\{x + 1, y + 1\}$ with two absent adjacent subcells will also be absent $q_a^{0|00}$; this yields, $p_{4a}^0 = p_a^0 \left(q_a^{0|0} \right)^2 q_a^{0|00}$. The calculation of the correlation between two adjacent empty cells can follow the same procedure, $q_{4a}^{0|0} = \left(q_a^{0|0} \right)^2 \left(q_a^{0|00} \right)^2$. Here, $q_a^{0|00}$ is the only probability unknown and can be estimated using the following Bayesian rule:

$$q_a^{0|00} = \frac{\left(q_a^{0|0} \right)^2 p_a^0}{\left(q_a^{0|0} \right)^2 p_a^0 + \left(q_a^{0|+} \right)^2 p_a^+}. \tag{1.10}$$

Consequently, we have the following occupancy and local density when the grain increases from a to $4a$ (Hui et al. 2006):

$$p_{4a}^+ = 1 - \frac{\nabla^4}{\Delta}$$

$$q_{4a}^{+|+} = \frac{\nabla^{10} - 2\nabla^4 \Delta^2 + \Delta^3}{\Delta^2 \left(\Delta - \nabla^4 \right)}, \tag{1.11}$$

where $\nabla = p_a^0 - q_a^{0|+} p_a^+$ and $\Delta = p_a^0 \left[1 - \left(p_a^+ \right)^2 \left(2q_a^{+|+} - 3 \right) + p_a^+ \left(\left(q_a^{+|+} \right)^2 - 3 \right) \right]$.
Evidently, a Poisson point pattern ($p_a^+ = q_a^{+|+}$) is the only scale-free spatial pattern, meaning $q_{4a}^{+|+} = p_{4a}^+$ and $p_{4a}^0 = \left(p_a^0 \right)^4$, with $p_a^0 = \exp(-\mu_a)$. Another proposition from this model is that the scaling pattern of occupancy is steepest when the local density $q^{+|+}$ is lowest. That is, when the spatial correlation between adjacent occupied cells is weak, the occupancy becomes more scale-dependent. As a cautious reminder, when scaling up, species distribution patterns change from random to

highly overdispersed (the value of σ^2/μ_a increases), whereas they changed from highly clustered to random (the value of $q_a^{+|+}/p_a^+$ declines). It is important to specify the meaning of 'aggregation' in communication.

The framework above can also be used to formulate the effect of grain shape on occupancy and spatial correlation. For instance, still combing four adjacent cells (but arranged along a transect with coordinates $\{x,y\}$, $\{x+1,y\}$, $\{x+2,y\}$ and $\{x+3,y\}$), we have $p_{4a}^0 = p_a^0 \left(q_a^{0|0}\right)^3$ and the conditional probability along the long edge $q_{4a}^{0|0} = q_a^{0|0}\left(q_a^{0|00}\right)^3$ and along the short edge $q_{4a}^{0|0} = \left(q_a^{0|0}\right)^4$. This can explain the edge effect that samples using more irregular shape (larger edge-to-size ratio) will overestimate occupancy and underestimate density, especially when individuals are spatially aggregated ($q_a^{+|+} > p_a^+$). Similar ideas have been used for designing efficient heat dissipation products, by increasing the surface/contact areas.

The value of developing the scaling pattern of occupancy lies in the potential for downscaling occupancy, that is, estimating occupancy at finer scales (higher resolution) based on observations at coarser scales (lower resolution). For downscaling, we could recursively solve p_a^+ and $q_a^{+|+}$ based on known p_{4a}^+ and $q_{4a}^{+|+}$ in Eq. (1.11), which only works for aggregated distributions ($q_a^{+|+} > p_a^+$). Alternatively, the model can be extended to the scenario of combining $n \times n$ unit-size cells. Let $a \equiv n \times n$, and we can have the following scaling pattern of occupancy and local spatial correlation (Hui 2009b):

$$p_a^+ = 1 - \alpha \cdot \beta^{2a^{1/2}} \gamma^a$$

$$q_a^{+|+} = p_a^+ + \frac{(1-p_a^+)^2}{p_a^+}\left(\alpha^{-1}\beta^{-a^{1/2}} - 1\right), \tag{1.12}$$

where $\alpha = p_1^0 q_1^{0|00}/\left(q_1^{0|0}\right)^2$, $\beta = q_1^{0|0}/q_1^{0|00}$ and $\gamma = q_1^{0|00}$ are three constants given a species' distribution. Downscaling can be applied using the curve-fitting method. Note, the above formulation, Eq.s (1.11) and (1.12), considers only correlations between adjacent cells and also deploys the Bayesian estimation. Extension can be developed but suffers from poor tractability. Nonetheless, the above scaling pattern of occupancy has withstood a number of tests and performed generally better than those based on specific probability distributions (e.g. Eqs. (1.2), (1.5) and (1.6); Hui et al. 2009).

The procedure above can be simplified by following a golden ratio bisection scheme for downscaling (Harte et al. 1999). In particular, the area has a width-to-height ratio of $\sqrt{2} : 1$, which is then subdivided into two similar rectangles with the same width-to-height ratio and so forth. The number of subareas is 2^i after the i number of subdivisions and the area of each one $A_i = A_0/2^i$, where A_0 is the size of the original area. Assuming self-similarity, if a species is present in an area, then there are two mutually exclusive states after one subdivision: both subdivided units are occupied with probability p, or only one subdivided unit is occupied, with probability q (note, $p + q = 1$). The expected value of the box-counting fractal dimension is thus $E(D) = 2 \ln (1 + p)/ \ln 2$. By induction, we can have the probability for the occupancy being $j/2^i$ ($j = 1, 2, 3, \ldots, 2^i$) after i bisections calculated in a recursive way (Hui and McGeoch 2007a):

$$P_i\left(\frac{j}{2^i}\right) = \sum_k C^k_{(j+k)/2} p^{\frac{i-k}{2}} q^k P_{i-1}\left(\frac{(j+k)/2}{2^{i-1}}\right), \tag{1.13}$$

where $k = 0, 2, 4, \ldots$ min $(j, 2^i - j)$ if j is an even number and $k = 1, 3, 5, \ldots$, min $(j, 2^i - j)$ if otherwise. When the self-similarity assumption breaks down, which often happens at finer scales, probability p and q need to be parameterised to become scale-dependent. In many cases, we have $p_i = 2c \cdot i^{-d} - 1$ and $q_i = 1 - p_i$ for $c^{1/d} \leq i \leq (2c)^{1/d}$; otherwise $p_i = 0$ for smaller i and 1 for larger i beyond the range (Hui and McGeoch 2007b).

1.5 Forecast Range Dynamics

Species range is not static, but dynamic. Forecasting the temporal trend of a focal species, its range expansion or retraction provides crucial information of population viability. Monitoring and forecasting population dynamics or trends often requires the accumulation of temporal records which is evidently time consuming. Forecasting distributional trends from a single snapshot of the current distribution could be an ideal method for conservation. By comparing the atlases of British butterflies calibrated from different periods, Wilson et al. (2004) discovered a strong correlation between the change in species range sizes and the exponent of the power-law scaling pattern of occupancy. This ability to forecast range dynamics from a single point pattern can be explained as the following (Hui 2011). Assuming colonisation only between adjacent cells and spatially random extinctions, the metapopulation dynamics can be depicted by the join-count statistics (Hui and Li 2004):

$$\dot{p}_a^+ = c \cdot p_a^{+0} - e \cdot p_a^+$$

$$\dot{p}_a^{++} = 2cp_a^{+0}\left(\frac{1}{z} + \frac{z-1}{z}q_a^{+|0+}\right) - 2e \cdot p_a^{++} \tag{1.14}$$

where c and e are the rate of colonisation and extinction, respectively, and z is the number of neighbouring cells that a coloniser can reach. $p_a^{+0} = p_a^+ - p_a^{++}$ and $p_a^{++} = p_a^+ q_a^{+|+}$ are the probabilities that a randomly chosen pair of neighbouring cells are in state +0 and ++, respectively (+ for presence and 0 for absence). The top equation implies that the colonisation only happens between a + 0 pair and that extinctions can occur randomly in any occupied cells. The bottom equation implies that a ++ pair can only emerge from the colonisation of a + 0 pair by the + cell in the pair or from a + 0 pair by the + neighbouring cells to the 0 cell in the pair and disappear from

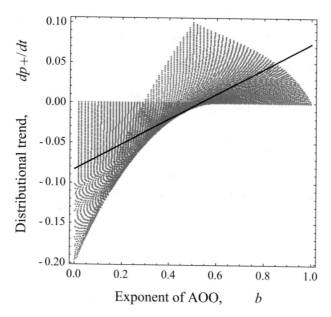

Fig. 1.2 The relationship between the exponent of area of occupancy (AOO) and the distributional trend. This is an illustration of the positive correlation between the AOO exponent and the distributional trend for the model depicted by Eq. (1.14), with 7000 points covering all feasible values of p_a^+ and $q_a^{+|+}$. The solid line indicates the result of the linear regression of these points ($c = 0.4$ and $e = 0.2$) (Redrawn from Hui (2011))

the extinction of any one of the + cell in the ++ pair (note, $p_a^{+0} = p_a^{0+}$ and thus the factor '2'). A ++ pair can also become a 00 pair but at the rate of e^2, and thus such events are ignored for simplicity. For closure, we also need to deploy pair approximation, $q_a^{+|0+} \approx q_a^{+|0}$. The solution to Eq. (1.14) is

$$\widehat{p}_a^+ = 1 - \frac{(z-1)e}{(z-1)c - e}$$
$$\widehat{q}_a^{+|+} = 1 - \frac{e}{c} \qquad (1.15)$$

Note $\widehat{p}_a^+ < \widehat{q}_a^{+|+}$, suggesting an aggregated spatial pattern. Occupancy p_a^+ is scale-dependent, and its scale dependency can be approximated by a power law, $p_a^+ \sim a^b$ (Kunin 1998), where the exponent can be estimated as $b = \ln \left(p_{4a}^+ / p_a^+ \right) / \ln (4)$ with p_{4a}^+ calculated using Eq. (1.11) or other similar ones described in previous sections. For the given spatial dynamics, we could assess the population dynamics $\dot{p}_a^+$ based on estimated exponent b from a single snapshot of spatial pattern (Fig. 1.2). This makes forecasting population dynamics possible based solely on a single spatial pattern. Note, changing model structures in Eq. 1.14 and parameters (c and e) will change the relationship between population trends and occupancy scaling.

1.6   Co-occurrence and Association

Species are not alone but associated positively or negatively (also known as spatial dissociation) with each other, due to, for instance, competition or facilitation in communities. Spatial association between two species is thus an important concept for understanding co-occurrence patterns in ecological communities – based on which, the type and strength of interspecific interactions, as well as community assembly processes, can be tentatively inferred. For two species, there exist four scenarios for a randomly chosen cell (Hui 2009b): species A and B coexist, p_a^{AB} (called the *joint occupancy*); only species A occurs, $p_a^{A\bar{B}}$; only species B occurs, $p_a^{\bar{A}B}$; and neither occurs, $p_a^{\bar{A}\bar{B}}$. Knowing the probability of one scenario, we can calculate the probabilities of the other three. For instance, knowing the joint occupancy p_a^{AB}, we have,

$$p_a^{A\bar{B}} = p_a^A - p_a^{AB}$$

$$p_a^{\bar{A}B} = p_a^B - p_a^{AB} \qquad (1.16)$$

$$p_a^{\bar{A}\bar{B}} = 1 - p_a^A - p_a^B + p_a^{AB},$$

where p_a^A and p_a^B are the occupancy of species A and B, respectively. We can define a positive association between species A and B as $p_a^{AB} > p_a^A p_a^B$, indicating a higher joint occupancy than expected when species A and B are independent from each other ($p_a^A p_a^B$). A negative association can be defined as $p_a^{AB} < p_a^A p_a^B$, and species independence as $p_a^{AB} = p_a^A p_a^B$. This definition of species association and co-occurrence is closely linked to the C-score null model test to examine species co-distributions (Gotelli and Graves 1996; Stone and Roberts 1990).

Species association is also scale-dependent. Let us first calculate the probability that both species are absent in the combined $4a$-size cell $p_{4a}^{\bar{A}\bar{B}}$. It equals the probability of neither species occurring in a subcell $p_a^{\bar{A}\bar{B}}$, multiplied by the conditional probability of species A also being absent in an adjacent subcell $q_a^{\bar{A}\|\bar{A}\bar{B}}$, multiplied by the probability that species B is also absent in this subcell $q_a^{\bar{B}|\bar{A}\|\bar{A}\bar{B}}$, multiplied by the probability of species B being absent in an adjacent subcell $q_a^{\bar{B}\|\bar{A}\bar{B}}$, multiplied by the probability that species A is also absent in this adjacent subcell $q_a^{\bar{A}|\bar{B}\|\bar{A}\bar{B}}$, multiplied by the probability of species A being absent in a chosen subcell – given that species A and B are also absent in its two adjacent subcells $q_a^{\bar{A}\|\bar{A}\bar{A}\bar{B}\bar{B}}$ – and finally multiplied by the probability that species B is absent in this subcell $q_a^{\bar{B}|\bar{A}\|\bar{A}\bar{A}\bar{B}\bar{B}}$. Based on similar pair approximation and Bayesian estimation, as in the calculation of Eqs. (1.11) and (1.12), we could explicitly formulate the above probabilities and derive the joint occupancy as

$$p_{4a}^{AB} = p_{4a}^{A} + p_{4a}^{B} - 1 + p_{4a}^{\bar{A}\bar{B}}, \tag{1.17}$$

The above equation provides a discrete description of species association. Analogous to Eq. (1.12), we can also have the probability when combining $n \times n \equiv a$ unit-size cells together in forming a larger cell, p_a^{AB}, as a continuous version of the scaling pattern of species association (Hui 2009b).

This formulation of co-occurrence has a few propositions. First, species independence is scale-free, regardless of the spatial structure (aggregation or segregation) of these two species. This is because we can have $p_{4a}^{AB} = p_{4a}^{A}p_{4a}^{B}$ under the sole condition of $p_a^{AB} = p_a^{A}p_a^{B}$. Second, the joint occupancy of species A and B, p^{AB}, has a nonlinear relationship when changing scales from a to $4a$. Even if $p_a^{AB} = 0$, the joint occupancy at larger scales will still become positive $p_{4a}^{AB} > 0$, indicating a higher chance of finding two species co-occurring at large scales, albeit with no overlaps at all at smaller scales. Based on the above formulation, we see that the category of species association, defined as being either positively or negatively associated, remains the same across scales. However, a positive association tends to be less positive ($p_a^{AB}/p_a^{A}p_a^{B}$ declines) when scaling up; similarly, a negative association tends to be less negative when scaling up. Therefore, the intensity of species association (i.e. the non-random co-occurrence) weakens with increasing grain size.

1.7 Species-By-Site Matrix

A community is composed of multiple resident species, and we can glimpse its biodiversity through sampling at multiple sites (at the same time or not). As a standard practice, original data from community ecology is often recorded as a species-by-site matrix, $M_{S \times N}$, with rows indicating species, columns different sites and the element the number of individuals (or incidence [i.e. presence/absence]) of a species in a specific site (see below for an example):

$$
\begin{array}{ccccc}
 & \text{site 1} & \cdots & \text{site } N & \text{row sums} \\
\text{species 1} & \begin{bmatrix} 2 & \cdots & 1 \end{bmatrix} & & & 7 \\
\vdots & \begin{bmatrix} \vdots & \ddots & \vdots \end{bmatrix} & & & \vdots \\
\text{species S} & \begin{bmatrix} 0 & \cdots & 3 \end{bmatrix} & & & 18 \\
\text{column sums} & 5 & \cdots & 10 & 135
\end{array}
$$

Previous sections essentially address issues related to one particular row (species occupancy) or two particular rows (species association). A similar matrix can also be used for describing a meta-community, where many local communities are connected via the dispersal of individuals, with one site in the matrix above representing a specific local community. Such a species-by-site matrix can be used for constructing multiple macroecological and community assemblage patterns.

Specifically, patterns depicting the relationship between rows (R-mode analysis) are mainly used for detecting interactions and associations between species, while patterns on the relationships between columns (Q-mode analysis) are used for sorting habitat hierarchies and compositional clusters (Legendre and Legendre 1998; Arita et al. 2008; Hui et al. 2013). For instance, the binary format of the matrix (also called the incidence matrix) has been widely used for calculating the co-occurrence patterns between species pairs, with observed patterns compared with those from null models for testing the signal of competition in communities (Gotelli and McCabe 2002).

For a species-by-site matrix with its elements representing abundance counts, the frequency or probability distribution of row sums represents the best-studied pattern in macroecology and community ecology – species abundance distribution (SAD) – also called the relative abundance distribution. Empirical studies have shown most SADs to conform to a right-skewed lognormal-like distribution, indicating many rare species but only a few abundant ones in a community. Many models have been developed to explain the emergence of this pattern. For instance, an SAD could reflect a distribution that maximises community entropy under different constraints (Harte 2011). As shown in Eqs. (1.8) and (1.9) where we partitioned N individuals into M cells, we could consider a scenario where S number of species tags are assigned to N individuals (i.e. partition N individuals into S species). The maximum entropy solution remains the same as Eq. (1.9). We could also consider constraints that are more specific. For instance, MacArthur's (1957) broken stick model assumes that the S number of resident species randomly partitions the available niche; this can be done by randomly drawing $S - 1$ numbers between 1 and N and ordering them from small to large, M_i ($i = 1, 2, \ldots, S - 1$). The abundance of species i can then be calculated as $N_i = M_i - M_{i-1}$, with $M_0 = 0$ and $M_S = N$. Tokeshi's (1990) dominance pre-emption model assumes that there is a clear dominance ranking among the resident species. The top species has a number of individuals N_1 randomly drawn between 1 and $p \cdot N$, the second species a number of individuals N_2 randomly drawn between 1 and $p(N - N_1)$, the third randomly drawn between 1 and p $(N - N_1 - N_2)$ and so on. The neutral theory of community is a dynamical model based on a crucial assumption of ecologically identical (or rather equivalent) species (Hubbell 2001). The stochastic niche model is also a dynamical model based on niche partitioning and stochastic population dynamics (Tilman 2004). Later chapters also introduce standard Lotka-Volterra models with interaction strengths randomly assigned or trait-dependent, which could also produce right-skewed SADs.

The other typical matrices in ecology include species-by-trait matrix, $T_{S \times Y}$, and the site-by-environment matrix, $E_{N \times K}$. Note, traits can be functional (e.g. morphological or physiological; body size, metabolic rate) or genetic (e.g. alleles at particular genetic markers). Environmental variables can include any site-specific measurements (e.g. elevation, temperature and rainfall, as well as geographical coordinates). These three matrices can also be temporally dynamic and allow us to explore questions related to ecological and evolutionary dynamics (see other chapters). With only these three matrices, we could visualise and explore their correlative relationships based on matrix manipulation and multivariate statistics (e.g. clustering and ordination based on ecological distances between sites or between species) (Legendre and Legendre 1998). For instance, $(MM^T)_{S \times S}$ is a symmetric matrix representing the ecological similarity between species based on their compositions (of abundance or incidence) in sites; $(M^T M)_{N \times N}$ is also a symmetric matrix representing the ecological similarity between sites based on species compositions. Similarly, $(TT^T)_{S \times S}$ is a matrix representing the trait similarity between species; $(EE^T)_{N \times N}$ is a matrix representing the environmental similarity between sites. We could also simply derive the trait-environment association, known as the fourth-corner problem (Brown et al. 2014), as $(T^T ME)_{Y \times K}$, with its element indicating the association between a particular trait and an environmental variable filtered through the species-by-site matrix.

1.8 Occupancy Frequency and Ranking

In a binary species-by-site matrix, the pattern depicting the frequency distribution of row sums (defined as the occupancy of a species) has been coined the occupancy frequency distribution (OFD) (McGeoch and Gaston 2002). Raunkiaer's (1934) law of frequency portrays a bimodal OFD of plant communities, suggesting species in a community are either rare or common, with few species having intermediate occupancies. Three explanations of the bimodality in OFDs exist: first, it represents an artefact from the sampling of highly skewed SAD. Repeated sampling with replacement from a community with either a truncated lognormal or log-series SAD will lead to a bimodal OFD (Papp and Izsák 1997). Second, the core-satellite hypothesis states that if local extinctions are subject to a strong rescue effect in a meta-community, the occupancy dynamics will lead to a bimodal OFD (Hanski 1982). Finally, the bimodality in OFDs could represent a transient pattern due to the effect of spatial scales on species occupancy (Hui and McGeoch 2007a, b, 2008). Reducing the sampling grain a will shrink species occupancies and thus lift up the frequency of rare species but push down the frequency of common ones. Intermediate grain sizes thus generate two modes in an OFD.

The parametric form of an OFD can be diverse, although its formulae are not well established, as typical probability distributions have only a single mode. In practice, we could instead explore the parametric form of occupancy ranking curve (ORC). Let $O(R)$ be the occupancy of a species with a ranking of R, $R(O)$ be the ranking of a species with occupancy O and $F(O)$ be the number of species (i.e. the frequency) of occupancy O in the community. Evidently, the rank of a species is equal to the number of species with greater occupancies than this species (Hui 2012):

$$R(O) = \int_{x=O}^{N} F(x)dx = \sum_{x=O+1}^{N} F(x), \qquad (1.18)$$

where N is the total number of sites (i.e. the maximum occupancy). Therefore, we have the derivative of the species' rank with respect to its occupancy, $\dot{R}(O) = -F(O)$. Since $R(O)$ is the inverse function of $O(R)$, we have $\dot{R}(O) = 1/\dot{O}(R)$, and thus the derivative of the occupancy ranking curve can be expressed as the negative reciprocal of the occupancy frequency (Hui 2012),

$$F(O) = -1/\dot{O}(R). \tag{1.19}$$

This means that OFD and ORC are mathematically transferable. The above equation provides a general relationship between rank curves and frequency distributions; for instance, we could also transfer the SAD to the abundance rank curve (another well-explored pattern in community ecology). To explore the bimodal OFD, we can further have $\ddot{O}(R) = -\dot{F}(O)/F(O(R))^3$, indicating that a bimodal OFD ($\dot{F}(\widehat{O}) = 0$ at the occupancy valley $\widehat{O}$) is mathematically equivalent to the existence of an inflection point $\ddot{O}(\widehat{R}) = 0$ at the ranking $\widehat{R}$. The dominant parametric form of ORC is the exponential power function, also known as the truncated power law (Hui 2012):

$$O(R) = a \cdot R^b \cdot e^{-c \cdot R}. \tag{1.20}$$

This form consists of two parts: a power-law function and an exponential cut-off. Note, this form can also possess an inflection point ($\widehat{O} = a\left(\left(\sqrt{b} + b\right)/c\right)^b e^{-\sqrt{b}-b}$ if $b > 0$). About one quarter of all communities have bimodal OFDs. Interestingly, a similar dominant form has also been observed for the node degree distribution in complex networks (see Chap. 4).

1.9   Biodiversity Partitioning and Scaling

The vector of row sums of a species-by-site matrix provides all the information necessary for expressing compositional biodiversity. Many biodiversity metrics have been developed to capture the feature of this vector. Of course, with trait and genetic information, these biodiversity metrics can be further modified to represent functional, genetic and phylogenetic diversity. We here specify only compositional diversity. The most well-known metric is the Shannon (1948) diversity index, Eq. (1.8), with p_i represents the relative abundance (i.e. the proportion of individuals) of species i. Another well-known metric is the Simpson (1949) index of evenness, $H_{Sim} = \sum p_i^2$. A more comprehensive metric is known as the Hill (1973) numbers,

$$H_\alpha = \left(\sum_{i=1}^{S} p_i^\alpha\right)^{1/(1-\alpha)}, \tag{1.21}$$

For $\alpha = 0$, the Hill number is simply the species richness, $H_0 = S$. When α approaches 1, the Hill number converges to the exponential of the Shannon-Wiener entropy index, $H_1 = \exp(H)$. For $\alpha = 2$, the Hill number becomes the reciprocal of the Simpson index, $H_2 = 1/H_{\text{sim}}$. Consequently, the Hill numbers are also called the effective numbers of species or species equivalents. When $\alpha \to +\infty$, the Hill numbers converge to the reciprocal of dominance, $H_{+\infty} = 1/\max(p_i)$. When $\alpha \to -\infty$, the Hill numbers converge to the number of rarest species needed to fill the community, $H_{-\infty} = 1/\min(p_i)$. With the increase of α from zero, the Hill numbers are gradually giving more weights to more abundant species. Because abundant species are infrequent in a community, H_α declines with increasing α.

When pooling samples (either by individual or by site), the number of observed species will increase, due to multiple reasons. This is called the species accumulation curve (Ugland et al. 2003). Assuming the probability of species i occurring in a sample is p_i^+ and that there is no interspecific interference, we have the expected number of species in m number samples, also known as the rarefaction curve,

$$E(S_m) = \sum_{i=1}^{S} \left(1 - \left(1 - p_i^+\right)^m\right). \tag{1.22}$$

Note, the probability of occurrence p_i^+ can be further implemented with specific models in Sect. 1.2. When samples are spatially distributed so that pooling samples represent increasing sampling extent, we have the species-area relationship, which normally resembles a power-law-like saturation curve.

Spatial variation in the presence or absence of species among sites, or compositional diversity of species turnover, is commonly quantified by beta diversity. The concept of beta diversity is derived from partitioning regional gamma (γ) diversity into alpha (α) and beta (β) components using either Whittaker's (1960) multiplicative ($\beta = \gamma/\alpha$) or Lande's (1996) additive ($\gamma = \alpha + \beta$) diversity partitioning. Existing measures of compositional similarity and dissimilarity are largely derived for pairwise comparisons of individual assemblages (sites, samples or areas) (Jost et al. 2011; McGlinn and Hurlbert 2012). Many pairwise dissimilarity metrics can depict the ecological distance between two sites, and thus the distance matrix between multiple sites is often used for statistical clustering and ordination (Faith et al. 1987; Legendre and Legendre 1998). Compositional similarity between two sites often declines with increasing distance between the two sites, known as the distance decay of similarity (Fig. 1.3). Distance decay relationships are valuable for estimating the rate of species turnover with distance and the importance of dispersal in driving the similarity of species assemblages at various scales (Qian and Ricklefs 2012).

Multisite compositional similarity metrics are also available, and hereafter we introduce the framework of zeta diversity (Hui and McGeoch 2014). Let the zeta component, ζ_i, be the number of species shared by i number of specified sites. Note that ζ_1 (where $i = 1$) is simply the mean number of species across all sites. Zeta diversity can of course be used to calculate the range of existing incidence-based, pairwise beta diversity and multiple-assemblage metrics. For example, Jaccard's (1900) similarity index is $\zeta_2/(2\zeta_1 - \zeta_2)$ and Sørensen's (1948) index is ζ_2/ζ_1. For

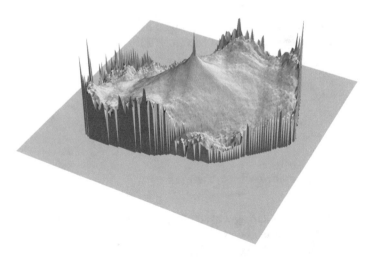

Fig. 1.3 Distance decay of similarity for Southern African birds. The anisotropic decline of the Jaccard compositional similarity between two quarter-degree cells with the increase of their geographic distance. Overlapping cells (with zero distance apart) are presented in the centre (Based on data from Hui et al. (2009))

multiple-assemblage similarity metrics, Koch's (1957) index of dispersity (i.e. taxonomic homogeneity) is $(\zeta_1/S_m - 1/m)/(1 - 1/m)$ and Diserud and Ødegaard's (2007) index is $(m - S_m/\zeta_1)/(m - 1)$. Since species shared by i sites will necessarily be among those shared by $i - 1$ sites, the number of shared species ζ_i declines monotonically with i (Hui and McGeoch 2014; Latombe et al. 2018):

$$
\begin{aligned}
E(\zeta_i) &= \sum_{j}^{S} E\left(p_j^i\right) \\
\mathrm{var}(\zeta_i) &= \sum_{j}^{S}\sum_{k}^{S} \mathrm{cov}\left(p_j^i, p_k^i\right)
\end{aligned}
\tag{1.23}
$$

where p_j^i is the probability of species j occurring in the specified i sites and can be simply calculated as $E\left(p_j^i\right) = C_{n_j}^i/C_N^i$, with n_j the occupancy (or incidence) of species j and N the total number of sites. The covariance between the probabilities of species j and k co-occurring in these i sites can be calculated as $\mathrm{cov}\left(p_j^i, p_k^i\right) = E\left(p_j^i p_k^i\right) - E\left(p_j^i\right)E(p_k^i)$, where $E\left(p_j^i p_k^i\right) = C_{n_{jk}}^i/C_N^i$ with n_{jk} the joint occupancy of species j and k (the corresponding element of MM^T for the incidence matrix M).

Zeta diversity (multisite similarity) can be used as a common currency to build all existing patterns in macroecology and community ecology. Let S_m be the total number of species across m sites, $F_{i,m}$ the number of species that occupy i sites out of the total m sites surveyed and $E_{i,m}$ the number of *locally* endemic species that only occur in the selected i sites, and we could have the following SAC, OFD and endemic-effort relationship (Hui and McGeoch 2014),

$$S_m = \sum_{k=1}^{m} (-1)^{k+1} C_m^k \zeta_k$$

$$F_{i,m} = C_m^i \sum_{k=1}^{m-i+1} (-1)^{k+1} C_{m-i}^{k-1} \zeta_{i+k-1} \quad (1.24)$$

$$E_{i,m} = \sum_{k=m-i+1}^{m} F_{1,k}/k$$

Adding one extra site to a survey will add $E_{1, m+1} = F_{1, m+1}/(m+1)$ number of new species. This means that the number of species in a region can be estimated when m approaches infinity (Hui and McGeoch 2014),

$$S_\infty = S_m + \sum_{k=m+1}^{\infty} F_{1,k}/k. \quad (1.25)$$

Evidently, a limited asymptote of observed species with increasing sample sites depends on the convergence of series $F_{1, k}/k$. When $\zeta_i = a \cdot \exp(-b \cdot i)$, the above formula can be simplified as $S_\infty = S_m + (n-1)F_{1,m}^2/(2nF_{2,m})$, known as the Chao (1984) II estimator, a lower bound estimator of species richness in a community. In most communities, zeta diversity declines much more slowly with i than the exponential form, with the most common two-parameter form a power law. The most common three-parameter form of the decline of zeta diversity with increasing order i is the exponential power law (replacing R with i in Eq. (1.20)).

Compositional diversity can be partitioned hierarchically by pooling or grouping sites into clusters or larger samples (Crist and Veech 2006). When pooling a number samples to form n ($m = n \times a$) larger grain clusters, the general form of zeta diversity is (Hui and McGeoch 2014):

$$\zeta_n(a) = \sum_{k=n}^{m} \frac{\sum_{x_j \geq 1, \sum x_j = k} \prod_{j=1}^{n} C_a^{x_j}}{C_m^k} F_{k,m}. \quad (1.26)$$

Accordingly, when pooling m_1 samples and another m_2 samples to form two sample clusters, the number of shared species between the two clusters is (Hui and McGeoch 2014):

$$\zeta_2(m_1, m_2) = \sum_{k=2}^{m_1+m_2} \frac{\sum_{i=1}^{k-1} C_{m_1}^i C_{m_2}^{k-i}}{C_{m_1+m_2}^k} F_{k, m_1+m_2}. \tag{1.27}$$

The distance decay of similarity has also been formulated using the zeta diversity framework (Hui and McGeoch 2014). In addition, once environmental variables are available at different sites (i.e. the site-by-environment matrix, $E_{N \times K}$), we could conduct pairwise or multisite generalised dissimilarity modelling to discern environmental drivers (including distance and temporal elapse between samples) that could have driven the observed species turnover pattern (Ferrier et al. 2007; Latombe et al. 2017).

1.10   Imperfect Detection and Sampling Effect

Ecological sampling normally cannot cover the entire range of a landscape due to limited efforts. As such, many individuals will not be recorded. When we transfer such a point pattern into presence-absence grids, we are essentially dealing with imperfectly detected individuals. Cells without a single individual recorded do not necessarily reflect the absence of the species but could merely indicate no detection given the sampling effort, even though the species might still in reality be present in the cell. This pseudo-absence dilemma can be solved in two ways: maximum likelihood and Bayesian methods. The maximum likelihood method combines two binomial processes in forming a joint likelihood for the probability of absence and detection for each cell. This can then be estimated by maximising the logarithmic likelihood by assuming a constant point detection rate (MacKenzie et al. 2003) or a predefined distribution of species abundance (Royle and Nichols 2003). The Bayesian method has also been applied to estimating abundance and occurrence in repeated surveys (Royle and Dorazio 2008).

For a landscape of size A with a number of n reported records (including x number of presences), if each record only corresponds to one non-repetitive sampling visit of an area of a with perfect detection, then a number of $M = A/a$ records are needed for obtaining the full information of this landscape. We assume that there are actually N presences (i.e. the true underlying abundance) once we had the full information of the M records. The detection rate of one random sample in this landscape thus equals N/M (i.e. the detection rate reflects the true abundance of a species in the landscape), which differs from the sampling incidence, x/n. The probability of finding x presences in the n reported records, knowing that there are N presences in the full information of M records, follows a hypergeometric distribution (Hui et al. 2011a):

$$\text{prob}(x|N) = C_N^x C_{M-N}^{n-x} / C_M^n. \tag{1.28}$$

Therefore, the probability distribution for true abundance N, given that x presences have been reported in the n samples, can be estimated by the Bayesian rule (Hui et al. 2011a):

$$\text{prob}(N|x) = \frac{\text{prob}(x|N)\text{prob}(\cdot|N)}{\sum\limits_{y=x}^{M-n+x} \text{prob}(\cdot|y)\text{prob}(x|y)}, \qquad (1.29)$$

where $\text{prob}(\cdot|N)$ is a prior probability of N presences in the cell regardless of any sampling information. In practice, we could use the Poisson model for the true absence scenario, $\text{prob}(\cdot|N) = \exp(-d \cdot M)$ for $N = 0$ (this is to handle the zero-inflation problem), and the uninformative prior (Jaynes 1968) for a true presence scenario, $\text{prob}(\cdot|N) = (1 - \exp(-d \cdot M))/\left(N \sum_{z=1}^{M} 1/z\right)$ for $N \geq 1$. For instance, the probability of absence in the cell, given that all n records were absences, can be calculated as $\text{prob}(N = 0 | x = 0)$. For occupancy modelling, sampling occupancy ($p_a^+ \equiv 1 - p_a^0 = x/n$) is normally less than the true occupancy $(1 - \text{prob}(N = 0 | x = 0))$ due to pseudo-absences and thus needs to be corrected.

Sampling patterns can also alter the shape of true macroecological patterns, due to the difference between sampling occupancy and true occupancy, evident in the hypergeometric distribution of sampling effect, Eq. (1.28). The probability of encountering very rare species is near zero, rising with true occupancy in a sigmoid fashion to approach 1 for very common species. For large M, the hypergeometric discovery probability can be approximated by a continuous normal density function $\mathbf{N}(x|\mu, \sigma)$ with the mean $\mu = nN/M$ and standard deviation $\sigma = nN(1 - N/M)/M$. Let $f(x)$ be the number of species with the sampling occupancy x and $F(N)$ the number of species with the true occupancy N, that is, the true species richness in an area $S = \sum\limits_{N=1}^{M} F(N)$. As the sampling OFD $f(x)$ is known while the true OFD $F(N)$ is unknown, we have the inverse problem of solving the following Fredholm equation of the first kind (supplementary S2 in Kunin et al. 2018),

$$f(x) = \sum_{N=1}^{M} \text{prob}(x|N)F(N) \approx \int_{N=1}^{M} \mathbf{N}(x|\mu, \sigma)F(N)dN. \qquad (1.30)$$

In practice, we can first assume $F(N)$ follows a specific probability distribution (e.g. an informative prior or a lognormal distribution) and then fits the parameters of this probability distribution using least squares or other approaches.

An alternative method to handle the problems with imperfect detection and sampling effect is the tessellation polygon method. In particular, the convex hull (Fig. 1.4a) and the alpha hull have often been used for the estimation of the range extent and the area of occupancy, respectively. The convex hull was first introduced as a standardised way to handle haphazard point records (Rapoport 1982). It was further refined through the introduction of the alpha hull methodology,

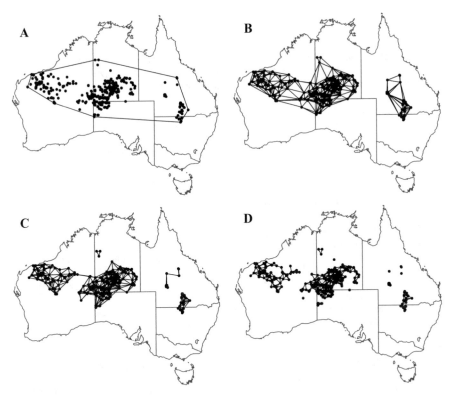

Fig. 1.4 Illustrations of the geographical range of *Acacia maitlandii* in Australia. Dots represent herbarium records of the species. (**a**) The convex hull; (**b**) the Delaunay triangulation; (**c**) and (**d**) alpha hulls at $d=256$ and 128 km (Redrawn from Hui et al. (2011b))

where the hull is split to exclude large uncertain areas without presence records. The alpha hull of a species can be calculated according to the following procedure (Burgman and Fox 2003). We first generate a Delaunay triangulation for the point pattern that maximises the minimum angles of all non-intersect triangles (Fig. 1.4b) and then remove edges longer than $\alpha \cdot L$, where L is the mean length of all edges (Fig. 1.4c, d). The total area of all remaining triangles is considered the area of occupancy for specific α values. Because the distribution of the points will affect the mean edge length L, we could refine this procedure by removing edges longer than a specific threshold (d). The scaling pattern of the alpha hull is described as (Fig. 1.4; Hui et al. 2011b):

$$P_d = P_c\left(1 - \exp\left(-a \cdot d^b\right)\right), \tag{1.31}$$

where the convex hull (P_c) of a point pattern is gradually filled up by an expanding alpha hull (P_d) when increasing the cut-off threshold. In this way, we could describe species occupancy, co-occurrence and other biodiversity patterns without the need to transfer a point pattern into a grid system.

References

Anselin L (1995) Local indicators of spatial association - LISA. Geogr Anal 27:93–115

Arita HT, Christen JA, Rodriguez P, Soberon J (2008) Species diversity and distribution in presence-absence matrices: mathematical relationships and biological implications. Am Nat 172:519–532

Brown AM, Warton DI, Andrew NR, Binns M, Cassis G, Gibb H (2014) The fourth-corner solution–using predictive models to understand how species traits interact with the environment. Methods Ecol Evol 5:344–352

Burgman MA, Fox JC (2003) Bias in species range estimates from minimum convex polygons: implications for conservation and options for improved planning. Anim Conserv 6:19–28

Chao A (1984) Nonparametric estimation of the number of classes in a population. Scand J Stat 11:265–270

Cliff AD, Ord JK (1981) Spatial processes: models and applications. Pion, London

Crist TO, Veech JA (2006) Additive partitioning of rarefaction curves and species–area relationships: unifying α-, β- and γ-diversity with sample size and habitat area. Ecol Lett 9:923–932

Diserud OH, Ødegaard F (2007) A multiple-site similarity measure. Biol Lett 3:20–22

Ellis EC, Ramankutty N (2008) Putting people in the map: anthropogenic biomes of the world. Front Ecol Environ 6:439–447

Faith DP, Minchin PR, Belbin L (1987) Compositional dissimilarity as a robust measure of ecological distance. Vegetatio 69:57–68

Ferrier S, Manion G, Elith J, Richardson K (2007) Using generalized dissimilarity modelling to analyse and predict patterns of beta diversity in regional biodiversity assessment. Divers Distrib 13:252–264

Gotelli NJ, Graves GR (1996) Null models in ecology. Smithsonian Institute Press, London

Gotelli NJ, McCabe DJ (2002) Species co-occurrence: a meta-analysis of J. M. Diamond's assembly rules model. Ecology 83:2091–2096

Hanski I (1982) Dynamics of regional distribution: the core and satellite species hypothesis. Oikos 38:210–221

Harte J (2011) Maximum entropy and ecology: a theory of abundance, distribution and energetics. Oxford University Press, Oxford

Harte J, Kinzig A, Green J (1999) Self-similarity in the distribution and abundance of species. Science 284:334–336

He F, Gaston KJ (2000) Estimating species abundance from occurrence. Am Nat 156:553–559

He F, Gaston KJ (2003) Occupancy, spatial variance, and the abundance of species. Am Nat 162:366–375

Hill MO (1973) Diversity and evenness: a unifying notation and its consequences. Ecology 54:427–432

Hubbell SP (2001) The unified neutral theory of biodiversity and biogeography. Princeton University Press, Princeton

Hui C (2009a) A Bayesian solution to the modifiable areal unit problem. In: Hassanien AE, Abraham A, Herrera F (eds) Foundations of computational intelligence, vol 2.: Approximate Reasoning. Springer, Berlin, pp 175–196

Hui C (2009b) On the scaling pattern of species spatial distribution and association. J Theor Biol 261:481–487

Hui C (2011) Forecasting population trend from the scaling pattern of occupancy. Ecol Model 222:442–446

Hui C (2012) Scale effect and bimodality in the frequency distribution of species occupancy. Community Ecol 13:30–35

Hui C, Li ZZ (2004) Distribution patterns of metapopulation determined by Allee effects. Popul Ecol 46:55–63

Hui C, McGeoch MA (2007a) A self-similarity model for the occupancy frequency distribution. Theor Popul Biol 71:61–70

Hui C, McGeoch MA (2007b) Modelling species distributions by breaking the assumption of self-similarity. Oikos 116:2097–2107

Hui C, McGeoch MA (2008) Does the self-similar species distribution model lead to unrealistic predictions? Ecology 89:2946–2952

Hui C, McGeoch MA (2014) Zeta diversity as a concept and metric that unifies incidence-based biodiversity patterns. Am Nat 184:684–694

Hui C, McGeoch MA, Warren M (2006) A spatially explicit approach to estimating species occupancy and spatial correlation. J Anim Ecol 75:140–147

Hui C, McGeoch MA, Reyers B, le Roux PC, Greve M, Chown SL (2009) Extrapolating population size from the occupancy-abundance relationship and the scaling pattern of occupancy. Ecol Appl 19:2038–2048

Hui C, Veldtman R, McGeoch MA (2010) Measures, perceptions and scaling patterns of aggregated species distributions. Ecography 33:95–102

Hui C, Foxcroft LC, Richardson DM, MacFadyen S (2011a) Defining optimal sampling effort for large-scale monitoring of invasive alien plants: a Bayesian method for estimating abundance and distribution. J Appl Ecol 48:768–776

Hui C, Richardson DM, Robertson MP, Wilson JRU, Yates CJ (2011b) Macroecology meets invasion ecology: linking the native distributions of Australian acacias to invasiveness. Divers Distrib 17:872–883

Hui C, Boonzaaier C, Boyero L (2012) Estimating changes in species abundance from occupancy and aggregation. Basic Appl Ecol 13:169–177

Hui C, Richardson DM, Pyšek P, Le Roux JJ, Kučera T, Jarošík V (2013) Increasing functional modularity with residence time in the co-distribution of native and introduced vascular plants. Nat Commun 4:2454

Jaccard P (1900) Contribution au proble`me de l'immigration postglaciaire de la flore alpine. Bull Soc Vaud Sci Nat 36:87–130

Jaynes ET (1968) Prior probabilities. IEEE Trans Syst Sci Cybern 4:227–241

Jost L, Chao A, Chazdon RL (2011) Compositional similarity and b (beta) diversity. In: Magurran AE, McGill BJ (eds) Biological diversity: frontiers in measurement and assessment. Oxford University Press, Oxford, UK, pp 66–84

Koch LF (1957) Index of biotal dispersity. Ecology 38:145–148

Kunin WE (1998) Extrapolating species abundance across spatial scales. Science 281:1513–1515

Kunin WE, Harte J, He F, Hui C, Jobe RT, Ostling A, Polce C, Šizling A, Smith AB, Smith K, Smart SM, Storch D, Tjørve E, Ugland KI, Ulrich W, Varma V (2018) Upscaling biodiversity: estimating the species-area relationship from small samples. Ecol Monogr 88:170–187

Lande R (1996) Statistics and partitioning of species diversity, and similarity among multiple communities. Oikos 76:5–13

Latombe G, McGeoch MA, Nipperess DA, Hui C (2018) zetadiv: Functions to compute compositional turnover using zeta diversity. R package, version 1.1.1, cran.r-project.org

Latombe G, Hui C, McGeoch MA (2017) Multi-site generalised dissimilarity modelling: using zeta diversity to differentiate drivers of turnover in rare and widespread species. Methods Ecol Evol 8:431–442

Legendre P, Legendre L (1998) Numerical ecology, 2nd edn. Elsevier, Amsterdam

Lloyd M (1967) Mean crowding. J Anim Ecol 36:1–30

MacArthur RH (1957) On the relative abundance of bird species. Proc Natl Acad Sci U S A 43:293–295

MacKenzie DI, Nichols JD, Hines JE, Knutson MG, Franklin AB (2003) Estimating site occupancy, colonization and local extinction probabilities when a species is detected imperfectly. Ecology 84:2200–2207

McGeoch MA, Gaston KJ (2002) Occupancy frequency distributions: patterns, artefacts and mechanisms. Biol Rev 77:311–331

McGlinn DJ, Hurlbert AH (2012) Scale dependence in species turnover reflects variance in species occupancy. Ecology 93:294–302

Millennium Ecosystem Assessment (2005) Ecosystems and human well-being: synthesis. Island Press, Washington DC

Mora C, Tittensor DP, Adl S, Simpson AGB, Worm B (2011) How many species are there on Earth and in the Ocean? PLoS Biol 9:e1001127

Moran PAP (1950) Notes on continuous stochastic phenomena. Biometrika 37:17–23

Morisita M (1962) Id-index, a measure of dispersion of individuals. Res Popul Ecol 4:1–7

Myers N, Mittermeier RA, Mittermeier CG, da Fonseca Gustavo AB, Kent J (2000) Biodiversity hotspots for conservation priorities. Nature 403:853–858

Olson DM, Dinerstein E (1998) The Global 200: a representation approach to conserving the Earth's most biologically valuable ecoregions. Conserv Biol 12:502–515

Openshaw S (1984) The modifiable areal unit problem. GeoBooks, Norwich

Papp L, Izsák J (1997) Bimodality in occurrence classes: a direct consequence of lognormal or logarithmic series distribution of abundances – a numerical experimentation. Oikos 79:191–194

Park SY, Bera AK (2009) Maximum entropy autoregressive conditional heteroskedasticity model. J Econ 150:219–230

Peleg S, Werman M, Rom H (1989) A unified approach to the change of resolution: space and gray-level. IEEE Trans Pattern Anal Mach Intel 11:739–742

Perry JN (1995) Spatial analysis by distance indexes. J Anim Ecol 64:303–314

Qian H, Ricklefs RE (2012) Disentangling the effects of geographic distance and environmental dissimilarity on global patterns of species turnover. Glob Ecol Biogeogr 21:341–351

Rapoport EH (1982) Aerography. Permagon Press, Oxford, UK

Raunkiaer C (1934) The life forms of plants and statistical plant geography being the collected papers of C. Raunkiaer. Clarendon Press, Oxford

Ripley BD (1976) The second-order analysis of stationary point processes. J Appl Probab 13:255–266

Royle JA, Dorazio RM (2008) Hierarchical modelling and inference in ecology: the analysis of data from populations, metapopulations and communities. Academic Press, New York

Royle JA, Nichols JD (2003) Estimating abundance from repeated presence absence data or point counts. Ecology 84:777–790

Shannon CE (1948) A mathematical theory of communication. Bell Labs Tech J 27:379–423

Simpson EH (1949) Measurement of diversity. Nature 163:688

Sørensen T (1948) A method of establishing groups of equal amplitude in plant sociology based on similarity of species content and its application to analyses of the vegetation on Danish commons. Biologiske Skrifter 5:1–34

Stone L, Roberts A (1990) The checker board score and species distributions. Oecologia 85:74–79

Taylor LR (1961) Aggregation, variance and the mean. Nature 189:732–735

Tilman D (2004) Niche tradeoffs, neutrality, and community structure: a stochastic theory of resource competition, invasion, and community assembly. Proc Natl Acad Sci U S A 101:10854–10861

Tokeshi M (1990) Niche apportionment or random assortment: species abundance patterns revisited. J Anim Ecol 59:1129–1146

Ugland KI, Gray JS, Ellingsen KE (2003) The species–accumulation curve and estimation of species richness. J Anim Ecol 72:888–897

Weiss MC, Sousa FL, Mrnjavac N, Neukirchen S, Roettger M, Nelson-Sathi S, Martin WF (2016) The physiology and habitat of the last universal common ancestor. Nat Microbiol 1:16116

Whittaker RH (1960) Vegetation of the Siskiyou Mountains, Oregon and California. Ecol Monogr 30:279–338

Wilson RJ, Thomas CD, Fox R, Roy DB, Kunin WE (2004) Spatial patterns in species distributions reveal biodiversity change. Nature 432:393–396

World Conservation Union (2014) IUCN Red List of Threatened Species, 2014.3. Summary Statistics for Globally Threatened Species. Table 1: Numbers of threatened species by major groups of organisms (1996–2014). International Union for Conservation of Nature, Switzerland

Wright DH (1991) Correlations between incidence and abundance are expected by chance. J Biogeogr 1:463–466

Chapter 2
Spread

Abstract Organisms move or disperse their progenies across space, either via their own motions or by currents and vectors. Spreading models differ on the level of tractability and realism. At the individual level, models of random walks have been developed to capture animal movement in heterogeneous landscapes. At the metapopulation level and regional scale, metapopulation models and demographic models have been developed to explore colonisation dynamics and spatial synchrony. When dispersal kernels are explicit, reaction-diffusion models and integro-difference equations, as well as agent-based models, can be applied for modelling the spread. The development of species distribution models, a set of sophisticated statistical tools, further allows us to trace the potential distribution of species in past and future environmental conditions.

2.1 Linear Birth-Death Process

Let us start with the simplest stochastic model of population dynamics, adding increasingly more realisms and complexities to models presented in later sections. Population dynamics can be considered an ensemble of birth and death events. Birth-death processes (BDP) are a class of Markov chains that model the number of individuals as a probabilistic (stochastic) variable. It essentially captures the effect of *demographic stochasticity* on population dynamics. At each time step, each individual can give birth to a number of individuals or die according to some probabilistic rules. Let $N(t)$ be the number of individuals at time t and $p_n(t) \equiv P[N(t) = n]$ be the probability that there are n individuals in the population during the short period of time from t to $t + \Delta t$. For a classic linear BDP, the only events that can occur during the short period Δt are the birth or death of one individual, with the probabilities of either happening for multiple individuals or both happening omitted, due to higher-order small chance events. Consequently, the probability $p_n(t + \Delta t)$ that there are n individuals in the population at time $t + \Delta t$ can be derived from three events:

© The Author(s), under exclusive licence to Springer International Publishing AG, part of Springer Nature 2018
C. Hui et al., *Ecological and Evolutionary Modelling*, SpringerBriefs in Ecology, https://doi.org/10.1007/978-3-319-92150-1_2

- There are $n - 1$ individuals at time t, and one new individual has been born during the Δt period.
- There are $n + 1$ individuals at time t, and one death has occurred to one individual during the period.
- There are n individuals at time t, and no individuals in the population have given birth or died during the period.

Once the probabilities of birth and death during the Δt period were set, we could simply simulate a linear BDP in an iteration procedure.

For a single birth to occur in a population with $n - 1$ individuals, it is necessary that exactly one individual gives birth while the other $n - 2$ individuals do not give birth. This can happen in $n - 1$ combinatorial ways. Let us assume that the probability of one individual giving birth during a small time period Δt is given by $\beta \Delta t + O(\Delta t)$, where $O(\Delta t)$ refers to a quantity approaching zero at a higher order of speed than Δt when Δt tends to be zero (i.e. the probability of more than one event to occur in the interval Δt is negligible). For simplicity, we drop $O(\Delta t)$ in the following formulation. We could write out the following probabilities for only one birth, or one death, to occur:

$$
\begin{aligned}
P[n\,|n - 1] &= (n - 1)(\beta\Delta t)(1 - \beta\Delta t)^{n-2} \\
P[n|n + 1] &= (n + 1)(\mu\Delta t)(1 - \mu\Delta t)^{n}
\end{aligned}
\tag{2.1}
$$

Consequently, considering only the first-order approximation in terms of Δt, we have the following recursive formulation of the probabilities: $p_n(t + \Delta t) = (n - 1)\beta\Delta t\, p_{n-1}(t) + (n + 1)\mu\Delta t\, p_{n+1}(t) + [1 - n(\beta + \mu)\Delta t]p_n(t)$. Rearranging the terms and taking the limit as $\Delta t \rightarrow 0$, we obtain the following differential equations for $n \geq 1$:

$$
\dot{p}_n = (n - 1)\beta\, p_{n-1} + (n + 1)\mu\, p_{n+1} - n(\beta + \mu)p_n,
\tag{2.2}
$$

with the boundary condition $\dot{p}_0 = \mu p_1$ and initial condition $p_n(0) = 1$ for $n = n_0$ and 0 otherwise. The system of equations (Eq. (2.2)) can be solved using probability or moment generating functions and techniques of partial differential equations (Kot 2001).

The statistical features of the above model can be summarised by the mean and variance of the population size:

$$
\begin{aligned}
\mathrm{E}[N(t)] &= n_0 e^{(\beta - \mu)t} \\
\mathrm{Var}[N(t)] &= n_0 \frac{\beta + \mu}{\beta - \mu} e^{(\beta - \mu)t}\left(e^{(\beta - \mu)t} - 1\right),
\end{aligned}
\tag{2.3}
$$

for $\beta \neq \mu$, and $\mathrm{Var}[N(t)] = 2n_0\beta t$ for $\beta = \mu$. For positive growth rate, defined as $r = \beta - \mu > 0$, both the mean population size and the variance increase exponentially. Moreover, when the initial population size is small, population dynamics can differ greatly from the expected population size in Eq. (2.3). The stochastic population is also subject to an extinction probability:

$$p_0(t) = \left(\frac{\mu \left(e^{(\beta - \mu)t} - 1 \right)}{\beta e^{(\beta - \mu)t} - \mu} \right)^{n_0}, \qquad (2.4)$$

for $\beta \neq \mu$, and $p_0(t) = (\beta t / (1 + \beta t))^{n_0}$ for $\beta = \mu$. Note, the extinction probability tends to 1 as $t \to \infty$ when $\beta = \mu$. When $\beta > \mu$, the extinction probability tends to $(\mu / \beta)^{n_0} < 1$, indicating that, albeit still probable, extinctions in large populations can be effectively evaded.

In a more general form, the probability distribution of the population follows this master equation:

$$\dot{P} = QP(t), \qquad (2.5)$$

where $P = (p_0, p_1, p_2, \ldots)^T$ is a vector containing the probability distribution of population size at time t and $Q = \langle q_{ij} \rangle$ is the transition matrix, with the elements describing the probability that the population size changes from i to j. In addition, the per capita birth and death rates are independent of population size in a linear BDP, i.e. being density-independent. However, density-dependent mechanisms do operate on birth, death or both. These density-dependent mechanisms can also be incorporated in the BDP.

Another common way to implement demographic stochasticity is to consider a finite number $N(t)$ of individuals at each time instant, each one characterised by its own fitness $f_i(t)$, that is, the number of offspring in the next time step. For instance, the individual fitness $f_i(t)$ can be 0, if the individual dies before reproduction; 1, if the individual survives without reproducing or produces one offspring and dies; 2, if the individual survives and produces one offspring or produces two offspring and dies, etc. The individual fitness $f_i(t)$ can then be extracted from a Poisson distribution, $P[f_i(t) = n] = \lambda^n e^{-\lambda}/n!$, where λ is the expected individual fitness. The average of the individual fitness across the population $f(t) = \sum_i f_i(t)/N(t)$ can then be used to compute the number of individuals in the next generation $N(t+1) = f(t)N(t) = \sum_i f_i(t)$. This dynamic might also sensibly differ from the expected dynamics $N(t+1) = \lambda N(t)$, especially for small populations.

2.2 Catastrophe and Stochasticity

Any types of BDP models can also be extended to consider environmental catastrophes and *environmental stochasticity*. A birth-death and catastrophe model implements the possibility of an arbitrary loss in the population size due to specified or unspecified external or environmental disturbance (Cairns and Pollett 2004). Let us assume a probability μ_k for a catastrophe of size k (death of k individuals) is acting over the population dynamics, where $\mu_k > 0$ for at least one $k \geq 1$. For instance, when $\mu_1 > 0$ and $\mu_k = 0$ for $k > 1$, we have a simple BDP (note, the death event now can be interpreted as caused by external force). In general, let f_n be the population change rate when there are n individuals in the population; let β be the probability of giving birth to a single individual. The catastrophe model can be described by a continuous-time Markov chain, with transition matrix:

$$\frac{q_{ij}}{f_i} = \begin{cases} -1 & j = i \geq 1 \\ \beta & j = i+1, i \geq 1, \\ \mu_{i-j} & j < i, i \geq 1 \end{cases} \qquad (2.6)$$

This population model can yield similar results to a linear BDP (Eq. (2.4)). In this model, the probability of extinction tends to 1 as $t \to \infty$ regardless of the initial population size if $\beta \leq \sum_k \mu_k$. If however $\beta \geq \sum_k \mu_k$, the extinction probability decreases with increasing initial population size. This model thus indicates the reality that the events of birth and death and the occurrence of different levels of catastrophes happen at different rates.

Environmental noise can affect the birth and death rates. The noise term is routinely formulated to capture the magnitude (noise variance), frequency and colour (noise autocorrelation) of environmental stochasticity. A symmetric logistic model, in which density-dependent processes operate on the birth and death rates, with environmental noise can be approximated by the Langevin stochastic differential equation:

$$\dot{n} = rn(1 - n/K) + \sigma_d \varepsilon_d(t)\sqrt{n} + \sigma_e \varepsilon_e(t)n, \qquad (2.7)$$

where n denotes the population size and K the carrying capacity (see logistic growth below), σ_e is the magnitude of the fluctuation in the population growth due to the environmental stochasticity captured by $\varepsilon_e(t)$ and σ_d is the intensity of fluctuations in the population growth rate due to demographic stochasticity (survivorship and reproductive success may differ between individuals) modelled by $\varepsilon_d(t)$. Note that $\varepsilon_d(t)$ and $\varepsilon_e(t)$ are independent. Intuitively, adding any kind of stochasticity in a model amplifies the fluctuation of population size and thus increases the extinction risk. The exact nature of the stochasticity however can yield a major surprise. In Eq. (2.7), the demographic noise is a white noise following a normal distribution, while the environmental noise is in general coloured. Analytical results on the trade-off between the magnitude and the frequency of the demographic or environmental noises are scarce, as most studies are based on simulations (but see Ovaskainen and Meerson 2010 and references therein). There is a common consensus however that slow environmental variations increase the extinction time more than rapid fluctuations.

2.3 Malthusian and Logistic Growth

Stochastic models can be informative, especially for a small population size, where environmental and demographical stochasticity could have great and profound impacts on population growth. However, stochastic models are normally difficult to analyse due to their complexity. When the population size is large and its variability relatively small, we could opt to use deterministic models for modelling population dynamics. The deterministic model for a linear BDP population is given by the Malthusian equation:

$$\dot{n} = rn, \tag{2.8}$$

which can be explicitly solved, $n(t) = n_0 e^{rt}$ with the initial population size being $n(0) = n_0$. The population increases or decreases exponentially at a rate determined by r, the intrinsic growth rate. This model, also known as the exponential model, is only accurate for a short period but nonetheless useful.

In reality, the per capita growth rate $\dot{n}/n$ is seldom constant but changes with many factors. For instance, when there are only limited resources to support growth, we need to reduce the per capita growth rate of an increasing population size. A simple adjustment is to discount the constant rate by multiplying the proportion of remaining resources for growth, $(K - n)/K$, with K representing the carrying capacity (i.e. the maximum environmental load of the population; Hui 2015). This discount factor is one of many density-dependent mechanisms that can regulate at different density levels. This yields the following logistic equation:

$$\dot{n} = \frac{rn(K - n)}{K}. \tag{2.9}$$

The exponential and logistic growth models are solvable analytically. For more complex models, different techniques can be used to investigate the dynamical behaviours of the model without the need to solve the model analytically (see Murray 2001).

Many species have discrete nonoverlapping generations (e.g. annual herbs and mayflies), often with reproduction and growth happening only at specific seasons. The population dynamics of such species can be modelled using difference equations:

$$n_{t+1} = f(n_t). \tag{2.10}$$

The density-independent case can be specified with $f(n) = \lambda n$, yielding the population dynamics, $n_t = n_0 \lambda^t$. This is similar to the exponential model with $r = \ln(\lambda)$. Different functions of $f(n)$ have been proposed to model the logistic growth for discrete-time populations. Among the commonly known is the Ricker model (Ricker 1954):

$$n_{t+1} = n_t \exp(r(1 - n_t/K)). \tag{2.11}$$

Discrete-time models can give rise to a wide spectrum of dynamical behaviours (Elaydi 2005; Brauer and Castillo-Chavez 2012), which often differ from the behaviours of their continuous-time counterparts (Fig. 2.1).

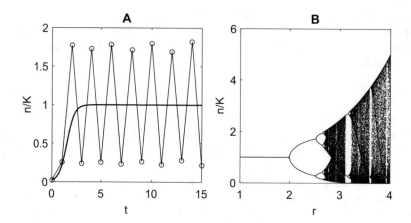

Fig. 2.1 Population dynamics of logistic growth. (**a**) Illustrations from continuous (Eq. (2.9), solid line) and discrete (Eq. (2.11), open circles) models, for $r = 2.6$. (**b**) Bifurcation diagram of the discrete model with respect to varying growth rate r. Points along any vertical lines represent population densities from $t = 125$ to 250. Other parameters: $K = 50$ and $n_0 = 1$

So far, we have considered population dynamics for a confined area. Organisms do move and disperse their progenies across spatial terrain, for many reasons. Mathematical models of spreading vary in their levels of detail. We introduce a few selected models in the sections to follow.

2.4 Metapopulation Dynamics

Metapopulation is a number of local populations where each local population is largely confined to a patch of suitable habitat with individuals of local populations moving between patches (Levins 1969). The classic model ignores detailed birth and death processes for each local population and focuses only on explaining the colonisation and extinction (i.e. presence or absence) of local populations in accessible habitat patches. The spreading dynamics are essentially interpreted as the dynamics of occupancy (i.e. the proportion of occupied patches, p). Let us assume that the rate of independent extinctions in local populations is e and that the colonisation rate of empty patches $(1 - p)$ by existing local populations (p) is c. We can then have the following Levins patch occupancy model:

$$\dot{p} = cp(1 - p) - ep. \tag{2.12}$$

It is equivalent to the logistic equation (Eq. (2.9)) with the growth rate $r = c - e$ and the carrying capacity $K = 1 - e/c$. That is, when $c > e$ the proportion of occupied patches will increase until it reaches $\hat{p} = K$. Note, although each local population suffers from certain extinctions at a constant rate, the persistence is assured at the metapopulation level; this emphasises two important concepts: dispersal and scale.

Metapopulation models have received great attention and development (Hanski 1999). For instance, the size and location (centrality) of a patch in the patch network could greatly affect the rate of extinction and recolonisation (van Nouhuys 2016).

2.5 Spatial Synchrony of Landscape Demography

Besides the metapopulation theory, a number of stochastic demography theories exist for multisite population viability analysis, which can be grouped under the banner of 'spatial synchrony of landscape demography'. Drivers of spatial synchrony between spatially differentiated populations can be depicted by the covariance of population sizes or vital rates. The Moran effect of broad-scale environmental regulation (Ripa and Ranta 2007), mediated interactions (e.g. shared natural enemy; Fox et al. 2011), density-independent (passive) dispersal (the case of classic metapopulation and source-sink dynamics; Dias 1996) and density-dependent dispersal (Li et al. 2005; Ramanantoanina et al. 2011) could all generate certain levels of co-varying population dynamics.

A number of theoretical frameworks have been developed to incorporate the effect of spatial synchrony and to explore modified population viability:

- Metapopulation theory: Population persistence at the regional or landscape scale is assured through the balance between (re)colonisation and extinction (Hanski 1999);
- Scale transition theory: The landscape-level multiplicative growth rate can be estimated as the summation of the average growth rate of local populations plus the growth-density covariance, and thus landscape-level growth can be elevated by positive growth-density covariance (Chesson et al. 2005);
- Growth inflation from pink noise: Landscape-level growth rate can be further inflated by temporal variability and autocorrelation in environmental noise, driving dynamics of intermittent rarity and facilitating population establishment (Roy et al. 2005; Gonzalez and Holt 2002);
- Statistic resonance: Landscape-level persistence can be achieved through connecting local stochastic processes via redistributing individuals, with the extreme case of a Parrondo game (Jansen and Yoshimura 1998; McDonnell and Abbott 2009);
- Landscape portfolio effect: Volatility reduction and growth inflation in landscape demography can occur from skewed distributions of co-varying local population growth rates, tilting the classic portfolio effect of volatility reduction to permitting growth inflation (Hui et al. 2017).

Let $\lambda = n_{t+1}/n_t$ be the multiplicative growth rate and $r = \ln(n_{t+1}/n_t)$ be the relative growth rate. Let $w = \langle w_i \rangle$ be the weight vector of populations in the landscape, called the *landscape portfolio*, with $w_i = n_i/\sum_i n_i$. Let $\mu = \langle \mu_i \rangle$ be the vector of expected local multiplicative population growth rate, with $\mu_i = E(\lambda_i)$, and $\mathbf{C} = \langle \sigma_{ij} \rangle$ be the covariance matrix of local multiplicative change rates, with $\sigma_{ij} = \text{cov}\,(\lambda_i, \lambda_j)$. The expected mean and variance of multiplicative population change rate for the ensemble of local populations can be estimated as follows, $\boldsymbol{\mu w}^{\mathrm{T}}$ and $\boldsymbol{w C w}^{\mathrm{T}}$, suggesting the reduction of volatility at the landscape level (Markowitz 1952). However, the distributions of population size and thus multiplicative growth rate are, in most cases, highly skewed, making the mean and variance of λ not valid metrics for the centroid and spread of population dynamics. To this end, we need to use the relative growth rate to describe population dynamics. The landscape-level relative growth rate can be calculated as the following (Hui et al. 2017):

$$R = \ln \left(\sum_i \exp(r_i) w_i \right). \tag{2.13}$$

The mean and variance can be further estimated, with the mean as follows:

$$E(R) = \ln \left(\alpha \sum_i \left(\text{cov}(\lambda_i, w_i) + \bar{w}_i \exp \left(\gamma_i + \frac{1 + \rho}{1 - \rho} \frac{\delta_i^2}{2} \right) \right) \right), \tag{2.14}$$

with $\gamma_i = E(r_i)$ and $\delta_i^2 = \text{Var}(r_i)$. This formulation includes all stochastic effects for growth inflation mentioned above: negative covariance between local population growth rates (summarised as α) for the portfolio effect, positive covariance $\text{cov}(\lambda_i, w_i)$ as in the scale transition theory, stochastic resonance $(\bar{w}_i)$ from joint random variables and positive temporal autocorrelation (pink noise; $0 < \rho < 1$ and variability (δ_i^2) of environmental noise; see Hui et al. (2017) for details).

2.6 Isotropic Random Walks

The standard approaches for modelling spread phenomena are those of random walk processes. The basis of random walk theory can be traced back to the irregular motion of individual pollen particles, known as the Brownian motion. Simple models of movement using random walks are simulating unbiased and uncorrelated motions. Unbiased here means isotropic; that is, there is an equal chance to move in any direction – while uncorrelated means the motion in the current step is independent of the motions in preceding steps. This is a Markov process with respect to the location of an individual. For instance, consider an individual walking on an infinite one-dimensional uniform lattice. Starting from the origin ($x = 0$), the individual moves a distance Δx either to the left or right during a short period of time Δt. Let d_i denote the displacement at each step. For an unbiased uncorrelated walk, after N steps, the position of the individual walker is $x(N) = \sum_{i=1}^{N} d_i$. It is obvious that with equal probabilities to move left and right the expected position $E(x(N))$ remains at the origin. The mean-squared displacement (MSD) from the origin is given by:

$$\text{MSD} = \sum_{i=1}^{N} d_i^2 = \Delta x^2 N. \qquad (2.15)$$

Let $t = N\Delta t$ and $x = m\Delta x$. The MSD of the individual at time t is given by:

$$\text{MSD}_t = 2Dt, \qquad (2.16)$$

where $D = \Delta x^2/(2\Delta t)$. Although the probability distribution of the location of the walker is still centred at the origin, its variation increases with time. Specifically, the probability of finding the particle at location x at time t follows a Gaussian distribution with time-dependent variance (Kot 2001):

$$P(x, t) = \exp\left(-x^2/(4Dt)\right)/\left(2\sqrt{\pi Dt}\right), \qquad (2.17)$$

where the parameter D is commonly referred to as the diffusion coefficient. More realistic scenarios, such as a directional bias or a waiting time before a dispersal event, as well as persistent and correlated walking directions, can be added to the simple isotropic random walk (Codling et al. 2008).

The MSD provides us with an average portrayal of coverage or diffusivity of the individual. However, we often also want to know the location of the furthest forward disperser (FFD) among a number of moving individuals, with or without interference between individuals (Reluga 2016). Let $n(t)$ denote the population size at time t of a species with nonoverlapping generations, with individual i located at x_i ($i = 1, \ldots, n(t)$) and that $x_1 < \ldots < x_{n(t)}$. Let $B_{i,\,t}$ represent the number of offspring of parent i at time t, further assuming no interference between parental individuals during the birth phase. During the dispersal phase, a dispersal distance $y_{t,\,i,\,j}$ for the j-th offspring of parent i is drawn from a dispersal kernel k. After the birth and dispersal phases, there are $\sum_{i=1}^{n(t)} B_{i,t}$ individuals in the system, located at $x_i + y_{t,\,i,\,j}$, with $i = 1, \ldots, n(t)$ and $j = 1, \ldots, B_{i,\,t}$ (Kot et al. 2004). During the competition phase, a density-dependent regulation of the population can be applied so that the minimal distance between two individuals is set to γ. This can be done by first identifying the two closest individuals and then removing the one on the right in the one-dimensional space, if their distance is less than the threshold γ. We can repeat this procedure until no two individuals are closer than distance γ. Note, a density-independent population model can be obtained by setting $\gamma = 0$. The competition process implies that the population size is bounded by

$$n(t + 1) \leq 1 + \left(x_{n(t)} - x_1\right)/\gamma. \qquad (2.18)$$

The distribution range of the population can be tracked by the change in the location of the FFD. Let $N_t = \left\{x_1^t, \ldots, x_{n(t)}^t\right\}$ and $N_{t+1} = \left\{x_1^{t+1}, \ldots, x_{n(t+1)}^{t+1}\right\}$ be the locations of individuals at time t and $t + 1$, respectively. We can consider the invasion step of the FFD as $z(t) = x_1^{t+1} - x_1^t$. The cumulative probability function of the invasion step is (Fig. 2.2a; Reluga 2016):

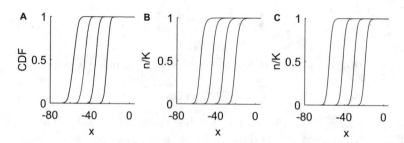

Fig. 2.2 An illustration of three spreading models. (**a**) Cumulative distribution of the position of the FFD and spatial distributions of population size from the PDE (**b**) and the IDE (**c**) models at time $t = 20, 30, 40, 50$. For the FFD model, we have $\beta = 2, \gamma = 0.1$. For the IDE and the PDE models, we have $r = \ln(2)$ and $K = 10$. The dispersal kernel for the FFD and IDE models follows the standard normal distribution. For the PDE model, $D = 0.5$

$$P(z|N_t) = 1 - \prod_{i=1}^{n(t+1)} B\big(1 - K\big(z - x_i^{t+1}\big)\big),\tag{2.19}$$

where $B(\cdot)$ is the probability generating function of the progeny numbers; that is $\$\$$ $B(s) = \sum_l r_l s^l$ where r_l is the probability that one individual has l offspring; $K(\cdot)$ is the cumulative probability density function of the dispersal kernel. Note, a random variable of Poisson distribution is normally used for modelling the progeny numbers, with its probability generating function given by $B(s) = e^{\beta s} e^{-\beta}$, with β the mean number of offspring per individual.

2.7 Diffusion and Travelling Waves

Classical methods to model the spread of a population in continuous time and space rely on reaction-diffusion partial differential equations. Let $n(x, t)$ be the number of individuals located in the interval $(x, x + \Delta x)$ at time t. The change of individual numbers in a small space interval Δx can be derived as follows:

$$\frac{\partial}{\partial x}[n(x, t)\Delta x] = G(x, t)\Delta x + J(x, t) - J(x + \Delta x, t),\tag{2.20}$$

where $G(x, t)$ is the local growth rate in a unit length and J the flux of individuals towards a direction of increasing x. For random diffusions, the flux is proportional to the negative density gradient; that is, individuals move from locations with a high population density to those of a low population density, known commonly as Fick's law. In a one-dimensional case we have:

$$J = -D\frac{\partial n}{\partial x}.\tag{2.21}$$

Consequently, the spatiotemporal dynamics of the population can be captured as follows (Fig. 2.2b):

$$\frac{\partial n}{\partial t} = g(n) + D\frac{\partial^2 n}{\partial x^2}, \tag{2.22}$$

where the growth rate depends implicitly on time and space through the population density, $G(x,t) = g(n(x,t))$.

To investigate the spread of populations, we normally explore the travelling wave solutions of the reaction-diffusion equation above. These are solutions of the form $n(x,t) = w(x - ct)$, indicating that $n(x,t)$ has a constant temporal wave profile w which propagates with a constant speed c. The wave profile satisfies the equation

$$\ddot{w} + c\dot{w} + g(w) = 0, \tag{2.23}$$

which is determined by the growth function. When the population follows an exponential growth function ($g(n) = rn$), the model is known as the Skellam model. When growth follows the logistic function ($g(n) = rn(1 - n)$), the model becomes the Fisher-KPP equation. For both cases, and more generally, when the growth rate is maximal at low population density, the population expands its range at a constant rate:

$$c = 2\sqrt{g'(0)D}. \tag{2.24}$$

Reaction-diffusion models and their travelling wave solutions have been extensively used in biological and ecological modelling to investigate the spread of a single or interacting species in a homogeneous environment (Shigesada and Kawasaki 1997; Murray 2003; Volpert and Petrovskii 2009).

2.8   Integro-Difference Equations

Diffusion as described above is but one specific type of motion. An alternative, more flexible, model for spreading dynamics is the integro-difference or integro-differential equations (IDEs). The IDE can implement any given shapes of dispersal or redistribution kernel, $k(x,y)$, that captures the probability density function of individual displacement, from location y moving to the interval $[x, x + dx]$. Although the differential models are used for continuous population growth, the integro-difference models are more appropriate for discrete-time population with nonoverlapping generations and separated growth and dispersal phases. With $g(\cdot)$ the function of growth, we could compute the population size at location x at time $t + 1$ as the total number of migrants from all locations y after the growth process (Fig. 2.2c):

$$n(x, t + 1) = \int k(x, y) g(y, n(y, t)) dy, \tag{2.25}$$

where $k(x, x)$ represents the probability that an individual remains in its natal place. For isotropic motion in a homogeneous landscape, the dispersal kernel can be simplified to depend solely on displacement distance, $k(x, y) = k(\|x - y\|)$, with the growth process only a function of local density, $g(y, n) = g(n)$. With such simplification, we can have the following travelling wave speed of spread if $g(n) < g'(0)n$ and the dispersal kernel exponentially bounded (Kot et al. 1996):

$$c = \sqrt{\sigma^2 \ln(g'(0))}, \tag{2.26}$$

where σ^2 is the variance of the dispersal kernel. Note that the rate of spread is the same as the one from the reaction-diffusion model (Eq. (2.24)). When the dispersal kernel is not exponentially bounded, however, the population advances its range at an accelerating rate (Kot et al. 1996).

2.9   Spatial Sorting of Mixed Populations

When a population is composed of individuals with different rates of growth and dispersal, we could observe the phenomenon of spatial sorting (Shine et al. 2011), where stronger dispersers are gradually accumulated at the advancing range front. Let us consider a population with N types of individuals; each type has a unique dispersal kernel, $k_i(x, y)$, and let $n_i(x, t)$ be the population size of type i at location x and time t. The variance σ_i^2 of the dispersal kernel k_i provides a scalar measure of dispersal ability for individuals of type i. The growth phase of type i individuals can be modelled by a non-negative function $g_i(n_1, \ldots, n_N)$ satisfying $g_i = 0$ if $n = 0$ (no new types) and $g_i \leq R_i n_i$ with $R_i = \partial g_i / \partial n_i|_{\mathbf{n} = 0}$ (negative density-dependent growth). For simplicity, we consider the case for no difference in growth rate between types, $R_i = R$. One example is the Ricker model, $g_i = n_i \exp(r (1 - n_i/K))$, with $n = \sum n_i$ and $r = \ln(R)$. Assuming perfect heritability (i.e. all offspring of type i individuals are of type i), we could have the following IDE for type i individuals (Ramanantoanina et al. 2014):

$$n_i(x, t + 1) = \int k_i(x - y) g_i(n_1, \ldots, n_N) dy. \tag{2.27}$$

The total population is given by,

$$n(x, t + 1) = \int k(x, y; t) g(u) dy, \tag{2.28}$$

where $k(x, y; t) = \sum_{i=1}^{N} k_i(\|x - y\|) p_i(y, t)$ is the expected dispersal kernel at time t, with $p_i(y, t)$ the proportion of type i individuals in location y, $p_i = n_i/n$. Note, the expected dispersal kernel is dependent on both time and population composition.

With more than two types of individuals, the model yields two distinct phases of spread: an accelerating instantaneous rate of spread, followed by a constant rate of spread. The instantaneous rate of spread during the acceleration phase is given by:

$$c^*(t) \cong \sqrt{2\,r\,\sigma^2(x^*(t))}, \tag{2.29}$$

where $\sigma^2(x^*(t)) = \sum_{i=1}^{n} \sigma_i^2 p_i(x^*(t))$ represents the mean dispersal ability at the advancing range front $x^*(t)$ at time t. Both the instantaneous and the average rates of spread approach the same asymptotic rate (Ramanantoanina et al. 2014):

$$c = \sqrt{2\,r\,\sigma_N^2} \left(1 + \frac{r}{12}\gamma_2\right), \tag{2.30}$$

where σ_N^2 is the maximal dispersal ability in the population and γ_2 the kurtosis of the corresponding dispersal kernel. The dispersal kernel of each type of individual in the population can also follow different distribution, although making the model difficult to track.

2.10 Across Heterogeneous Landscapes

Dispersal strategy is often dependent on the habitat quality. For instance, a generic pattern of 'good-stay, bad-disperse' strategy has been observed in common starlings based on ringing data (Hui et al. 2012). As such, we could further consider habitat heterogeneity when modelling the spread. In the reaction-diffusion models, we could specify diffusion rate and other vital rates as functions of habitat quality (i.e. spatial location):

$$\frac{\partial n}{\partial t} = r(x)n\left(1 - \frac{n}{K}\right) + D(x)\frac{\partial^2 n}{\partial x^2}. \tag{2.31}$$

For instance, in a landscape that is composed of good and bad habitat patches, we could assume r and D as stepwise functions of r_G or r_B and D_G or D_B, respectively (Shigesada et al. 1986). We could also consider the case where habitat quality varies sinusoidally in space (Kinezaki et al. 2006). While the stepwise habitat highlights the importance of the proportion and length of favourable habitats, the sinusoidal habitat can allow us to test growth-dispersal trade-off on the spread of populations. The IDE model can also be extended to include environmental heterogeneity in similar settings (Ramanantoanina and Hui 2016). For instance, we could assume that the growth rate and carrying capacity in the Ricker model to be dependent on the suitability of habitat, with $g(\cdot)$ in Eq. (2.25) specified as $n(y)\,\exp\,(r(y)n(y)(1 - n(y)/K(y)))$. The IDE models are often implemented over realistic habitats of different levels of suitability predicted from species distributions models (see below), forming a hybrid model of dynamic spread models over realistic landscapes (Roura-Pascual et al. 2009).

2.11   Species Distribution Models

Most spreading models presented above can be implemented over realistic land-scapes, known as hybrid models (Hui and Richardson 2017). It normally requires us to know how population demography and vital rates change across real landscapes. For instance, for a focal area ($x \in \Omega$), we need to have the spatially explicit intrinsic population growth rate $r(x)$ in the model of Eq. (2.31) to simulate the spreading dynamics. This can be achieved in two steps. First, we need to map all essential environmental variables that can strongly influence population demographies across the landscape. For instance, we first measure the temperature, $E(x)$, across the landscape, using either ground sensors or remote sensing via satellites and drones with the help of geographical information systems (GIS). Second, we need data on the response curves (also known as reaction norms), describing the performance of the species along specific environmental gradients (e.g. growth rate along a thermal gradient, $r(E)$). The performance can be measured in controlled environments, e.g. in a laboratory or greenhouse, i.e. where the germination and growth rate can be measured under different temperatures. This then allows us to project the perfor-mance as habitat suitability over the landscape, based on observed environmental maps, $r(E(x))$ (Kearny and Porter 2009; Morin and Lechowicz 2008). However, the response curves of some vital rates for long-lived organisms (e.g. the mortality of trees) or those that are density-dependent are difficult to quantify through experi-ments, not mentioning the mismatch between the environment in a controlled experiment versus the one experienced by the organism in the wild. For instance, the performance of a lizard in a thermal chamber cannot match the experience of an ever-changing microclimate in the wild, and the lizard could simply hide under trees or rocks to avoid overheating from basking in the sun.

To bypass the complication, we could use correlative SDMs that combine a set of environmental variables to explain the observed occurrence data and to project the potential range of a focal species. Again, we need to rely on full-range environmental maps, $E(x)$. In addition, through field survey we can obtain information of local abundances (or the presence and absence) of the focal species in a number of locations $\{n(x_1), \ldots, n(x_N)\}$. This allows us to use correlative statistics, such as logistic regression and the boosted regression tree (e.g. Roura-Pascual et al. 2011), to identify the relationships of species abundance (or presence/absence) as a function of explanatory environmental variables:

$$n(x_i) = \widehat{n}(x_i) + \varepsilon_i = f(E(x_i)) + \varepsilon_i, \qquad (2.32)$$

with the last term representing residuals between predictions and observations. Assuming that such a function also holds in other unsurveyed areas in the landscape, we could then map the potential distribution as $\widehat{n}(x) = f(E(x))$. Associated with the data format and multivariate nature of the statistics, a number of issues and tech-niques are available to address specific biases that often prey upon such data (e.g. pseudo-absences, uneven samples, multicollinearity, cross-validation). A great amount of excellent information is available (Franklin 2009; Elith and

Leathwick 2009; rspatial.org), and we encourage interested readers to look into these sources of information. To build hybrid models based on the projection $\hat{n}(x)$ from correlative SDMs, we need further assumptions to link the potential distribution $$ $\hat{n}(x)$ to related vital rates (e.g. $r(x)$ and $K(x)$) in spread models. Many of these assumptions are debatable: For instance, we could assume that the observed distribution is at its dynamical equilibrium; this allows us to consider $\hat{n}(x)$ a transformed $K(x)$. However, non-equilibrium dynamics and source-sink dynamics could obviously distort the relationship $f(E(x))$. Overall, the study of hybrid models of spreading over real landscapes remains a field at the forefront of ecological research.

References

Brauer F, Castillo-Chavez C (2012) Mathematical models in population biology and epidemiology. Springer, Berlin

Cairns B, Pollett PK (2004) Extinction times for a general birth, death and catastrophe process. J App Prob 4:1211–1218

Chesson P, Donahue MJ, Melbourne BA, Sears ALW (2005) Scale transition theory for understanding mechanisms in metacommunities. In: Holyoak M, Leibold MA, Holt RD (eds) Metacommunities: spatial dynamics and ecological communities. University of Chicago Press, Chicago, IL, pp 279–306

Codling EA, Plank MJ, Benhamou S (2008) Random walk models in biology. J Roy Soc Interface 5:813–834

Dias PC (1996) Sources and sinks in population biology. Trends Ecol Evol 11:326–330

Elaydi S (2005) An introduction to difference equations. Springer, Berlin

Elith J, Leathwick JR (2009) Species distribution models: ecological explanation and prediction across space and time. Ann Rev Ecol Evol Sys 40:677–697

Fox JW, Vasseur DA, Hausch S, Roberts J (2011) Phase locking, the Moran effect and distance decay of synchrony: experimental tests in a model system. Ecol Lett 14:163–168

Franklin J (2009) Mapping species distributions: spatial inference and prediction. Cambridge University Press, Cambridge

Gonzalez A, Holt RD (2002) The inflationary effects of environmental fluctuations in source-sink systems. Proc Natl Acad Sci U S A 99:14872–14877

Jansen VAA, Yoshimura J (1998) Populations can persist in an environment consisting of sink habitats only. Proc Natl Acad Sci U S A 95:3696–3698

Hanski I (1999) Metapopulation ecology. Oxford University Press, Oxford

Hui C (2015) Carrying capacity of the environment. In: Wright JD (ed) International encyclopedia of the social and behavioral sciences, vol 3, 2nd edn. Elsevier, Oxford, pp 155–160

Hui C, Richardson DM (2017) Invasion dynamics. Oxford University Press, Oxford

Hui C, Fox GA, Gurevitch J (2017) Scale-dependent portfolio effects explain growth inflation and volatility reduction in landscape demography. Proc Natl Acad Sci U S A 114:12507–12511

Hui C, Roura-Pascual N, Brotons N, Robinson RA, Evans KL (2012) Flexible dispersal strategies in native and non-native ranges: environmental quality and the 'good-stay, bad-disperse' rule. Ecography 35:1024–1032

Kearny M, Porter W (2009) Mechanistic niche modelling: combining physiological and spatial data to predict species' ranges. Ecol Lett 12:334–350

Kinezaki N, Kawasaki K, Shigesada N (2006) Spatial dynamics of invasion in sinusoidally varying environments. Popul Ecol 48:263–270

Kot M (2001) Elements of mathematical ecology. Cambridge University Press, Cambridge

Kot M, Lewis MA, van den Driessche P (1996) Dispersal data and the spread of invading organisms. Ecology 17:2027–2042

Kot M, Medlock J, Reluga TC, Walton DB (2004) Stochasticity, invasions and branching random walks. Theor Popul Biol 66:175–184

Levins R (1969) Some demographic and genetic consequences of environmental heterogeneity for biological control. Bull Ent Soc Am 15:237–240

Li ZZ, Gao M, Hui C, Han XZ, Shi HH (2005) Impact of predator pursuit and prey evasion on synchrony and spatial patterns in metapopulation. Ecol Model 185:245–254

Markowitz HM (1952) Portfolio selection. J Finance 7:77–91

McDonnell MD, Abbott D (2009) What is stochastic resonance? Definitions, misconceptions, debates, and its relevance to biology. PLoS Comput Biol 5:e1000348

Morin X, Lechowicz MJ (2008) Contemporary perspectives on the niche that can improve models of species range shifts under climate change. Biol Lett 4:573–576

Murray JD (2001) Mathematical biology: I. An introduction. Springer, Berlin

Murray JD (2003) Mathematical biology: II. Spatial models and biomedical applications. Springer, Berlin

Ovaskainen O, Meerson B (2010) Stochastic models of population extinction. Trends Ecol Evol 25:643–652

Ramanantoanina A, Hui C (2016) Formulating spread of species with habitat dependent growth and dispersal in heterogeneous landscapes. Math Biosci 275:51–56

Ramanantoanina A, Hui C, Ouhinou A (2011) Effects of density-dependent dispersal behaviours on the speed and spatial patterns of range expansion in predator-prey metapopulation. Ecol Model 222:3457–3650

Ramanantoanina A, Ouhinou A, Hui C (2014) Spatial assortment of mixed propagules explains the acceleration of range expansion. PLoS One 9:e103409

Reluga TC (2016) The importance of being atomic: ecological invasions as random walks instead of waves. Theor Popul Biol 112:157–169

Ricker WE (1954) Stock and recruitment. J Fish Res Board Can 11:559–623

Ripa J, Ranta E (2007) Biological filtering of correlated environments: towards a generalised Moran theorem. Oikos 116:783–792

Roura-Pascual N, Bas JM, Thuiller W, Hui C, Krug RM, Brotons L (2009) From introduction to equilibrium: reconstructing the invasive pathways of the argentine ant in a Mediterranean region. Glob Chang Biol 15:2101–2115

Roura-Pascual N, Hui C, Ikeda T, Leday G, Richardson DM, Carpintero S, Espadaler X, Gómez C, Guénard B, Hartley S, Krushelnycky P, Lester PJ, McGeoch MA, Menke SB, Pedersen JS, Pitt JP, Reyes J, Sanders NJ, Suarez AV, Touyama Y, Ward D, Ward PS, Worner SP (2011) Relative roles of climatic suitability and anthropogenic influence in determining the pattern of spread in a global invader. Proc Natl Acad Sci U S A 108:220–225

Roy M, Holt RD, Barfield M (2005) Temporal autocorrelation can enhance the persistence and abundance of metapopulations comprised of coupled sinks. Am Nat 166:246–261

Shigesada N, Kawasaki K, Teramoto E (1986) Travelling periodic waves in heterogeneous environments. Theor Popul Biol 30:143–160

Shigesada N, Kawasaki K (1997) Biological invasions: theory and practice. Oxford University Press, Oxford

Shine R, Brown GP, Phillips BL (2011) An evolutionary process that assembles phenotypes through space rather than through time. Proc Natl Acad Sci U S A 108:5708–5711

van Nouhuys S (2016) Metapopulation ecology. eLS. https://doi.org/10.1002/9780470015902.a0021905.pub2

Volpert V, Petrovskii S (2009) Reaction-diffusion waves in biology. Phys Life Rev 6:267–310

Chapter 3
Evolution

Abstract Ever since Darwin's inception of speciation via natural selection, scientists have started to develop models to capture evolutionary dynamics. Two types of models have emerged: those rooted in population genetics versus those modelling the evolutionary dynamics of phenotypic traits, often in the broader context of entangled biotic interactions among populations. An example of the first type is phylogenetic modelling, aiming at building phylogenies based on genetic affinity between organisms of different species or localities. Models of the second type combine analytic tools, such as optimality and game theory in population (and community) ecology, to provide a modelling framework for phenotypic evolution due largely to trait-mediated biotic interactions. This chapter focuses on the second type of evolutionary modelling. In particular, we introduce evolutionary optimality models, evolutionary game theory and adaptive dynamics, as well as evolutionary distribution modelling and the Price equation. These models allow us to explore a plethora of evolutionary dynamics, especially the formulation of endogenous (sympatric) diversification by adaptive dynamics, known as evolutionary branching.

3.1 Concepts of Evolution

Models of evolutionary dynamics implement evolutionary processes through natural selection which favours fitter organisms or eliminates those unfit. As such, fitness is undoubtedly the core concept in evolution. *Fitness* depicts the compound effect of survival and reproduction. Natural selection essentially acts as an environmental filter (in a broad sense) of the physiological and morphological characteristics of imbedded organisms: those that are better suited to a particular habitat will survive and contribute more to the next generation. Phenotypes are expressions of the discrete genetic material of the organisms (genotypes) through particular environment filters. These filters are largely of two groups: the abiotic environment, comprising all climatic, physical and chemical factors, and the biotic component, comprising all biotic interactions and feedbacks with other organisms of the same or other kind. The offspring, now facing different selection pressures due to changed environments,

C. Hui et al., *Ecological and Evolutionary Modelling*, SpringerBriefs in Ecology, https://doi.org/10.1007/978-3-319-92150-1_3

inherits the *phenotypic traits* of their parents but with some small modification due to mutation. These traits are the subject of selection, for or against, in the population.

Given a sufficient amount of time, nature could select the single fittest *resident* phenotype for one static environment, after which evolution ceases. This is because natural selection can act only as a filter of phenotypic variation in a population based on the differences in the relative fitness of each phenotype. When fitness lacks a gradient between phenotypes, natural selection loses its filtering effect. However, changes in the environment, even if minute, will select a different phenotype as the fittest. Phenotypic variation, the essential material for natural selection to act, could be introduced through the immigration of new organisms from other populations (gene flow) or maintained through recombination during sexual reproduction. Phenotypic variation also relies on the imperfection of reproduction from one generation to the next: Point *mutation*, mostly harmful, can occur infrequently in the genetic material at specific genetic loci for few individuals, possibly generating a new phenotype in the population. Mutant and resident phenotypes will once again pass through the selection filter, with their frequencies in the population changing according to their fitness. Those survived will become the new residents of the population, lurking for contests with new incoming phenotypes.

Traditionally, organisms with similar physiological and morphological traits (e.g. based on fossils) are recognised as one species. Theoretically, sexual organisms that can breed without reproduction barriers (i.e. they can produce viable offspring) are considered to belong to the same species. Nonetheless, consensus is yet to be reached for the concept of 'species'. For instance, the boundary of species for asexual organisms (e.g. bacteria) can be defined based on genetic similarity with a certain debatable threshold. Once the concept of species is tentatively set, the most important question in biology, championed by Darwin himself, emerges: How does a species evolve and diversify into new ones? In particular, how fast and under what conditions? There are two main mechanisms explaining the diversification of species: *allopatric* (and also parapatric) versus *sympatric* speciation. Allopatric speciation relies on the spatial segregation of species: A geographical barrier can separate two groups of initially similar organisms which undergo different selection forces in their separate habitats and eventually become two species after the emergence of a reproduction barrier. Parapatric speciation represents the formation of reproduction barriers along a continuous environmental gradient without notable geographical dispersal barriers. Sympatric speciation does not rely on environmental gradient or geographical barriers: It describes speciation driven by disruptive selection among individuals dwelling in the same habitat, where extreme phenotypic traits possess higher fitness than intermediate ones (Maynard Smith 1966). For example, specialisation from competition for resources is one common cause of disruptive selection and thus sympatric speciation.

A plethora of models and theories exists in the study of evolution. We here choose a few selected ones to illustrate the key concepts in building and analysing a mathematical model for capturing the evolutionary dynamics of functional traits. Although these selected approaches are by no means representative of the full spectrum of evolutionary models, they are nonetheless built on similar concepts and principles and analysed following similar procedures. Interested readers can branch out to other models and theories based on the understanding of these selected models. In particular, we focus here on fitness optimality, evolutionary game theory and adaptive dynamics, as well as the evolutionary distribution modelling and the Price equation.

3.2 Phenotypic Traits

Evolution acts through natural selection on the physiological and morphological characteristics of individuals, known as phenotypic traits. A phenotype is the realisation, in a particular environment, of the individual's genetic makeup, or genotype. Thus, most phenotypes are heritable from generation to generation but can be expressed differently in different or changing environments (note the concept of epigenetics). In sexual organisms, the offspring inherits half of its genetic material from each sex; as such, sexual reproduction is a common approach to maintain genetic diversity, then expressed by phenotypic variability. Traits are in a broad sense part of life-history strategies. Not only does selection act on phenotypic traits but also on life-history strategies (e.g. phenology and plasticity, theories on bet-hedging strategy that modifies fitness variance when facing uncertain conditions; Cohen 1966). When a phenotypic trait is expressed by only a few genes, we have discrete phenotypes (e.g. blood type and eye colour); when a phenotypic trait is determined by numerous genes, they can be considered to be continuous (e.g. body height and body mass). In the long run, genetic frequencies tend to equilibrate (the Hardy-Weinberg law); thus evolution ceases. Asexual organisms rely on rare mutations to keep breaking the Hardy-Weinberg equilibrium, by altering their genetic material and thus the phenotypic traits across generations. Mutations with beneficial fitness will be selected and potentially fixed in the population.

Due to physiological and energetic constraints, traits are not free to evolve indefinitely but are also correlated (e.g. allometry correlating traits to body size). Each value of a phenotypic trait has its specific energetic cost and thus pros and cons in fitness. With limited energy and resources to allocate, each individual faces *trade-offs* when expressing its phenotypic potential. For example, large body size can have an advantage in competitive or trophic interactions but could suffer from a slower development. Selection normally favours a specific direction for trait evolution under particular environmental conditions, known as *directional* selection. Traits under directional selection will continue to evolve until physiological constraints and fitness trade-offs eventually bring evolutionary dynamics to a halt. Directional selection does not necessarily lead to the proliferation of fitness and could sometimes result in Red Queen dynamics (Van Valen 1973), runaway selection (or drift) or evolutionary suicide (Gyllenberg and Parvinen 2001). Directional selection can generate diversity only under different environmental conditions or via the physical separation of organisms (i.e. allopatric speciation). Sympatric diversity can however be generated and maintained through *disruptive* selection. In the case of disruptive selection, biotic interactions (often density-dependent) favour the rare extreme phenotype over the common intermediate phenotype, allowing two different phenotypic variants to emerge and diverge in the same environment.

3.3   Optimality Models

One of the first models for phenotypic evolution is based on optimality theory. If fitness can be quantified as a function of trait values, then it can be maximised under certain constraints. Selection on some traits can be effectively captured by such optimality models. For example, the evolution of optimal clutch size in birds (Parker and Begon 1986) has been formulated according to trade-off between the number and size of offspring. A greater number of offspring (clutch size) ensures a higher chance for at least one to survive, while a bigger egg also gives the individual a higher chance of survival. Let f_0 be the basic fertility and s_0 the basic individual survival rate. With additional reproductive energy ready for allocation, we assume that the x portion is allocated to increasing the number of eggs, while the $1 - x$ is to increasing the egg size (thus improving survival). As a result, the realised fertility becomes $f_0 + fx$, with f the energy-to-fertility conversion rate; the realised survival rate becomes $s_0 + s(1 - x)$, with s the energy-to-survival conversion rate. The fitness of this particular energy-allocation strategy (or allocation trait, x) is the number of surviving offspring in the next generation:

$$F(x) = (fx + f_0)(s(1 - x) + s_0). \tag{3.1}$$

Evidently, different traits will have different fitness consequences. Evolution is supposed to maximise fitness. As such, the optimal trait x^* for reproductive allocation can be computed as the optimal value of x maximising fitness $F(x)$. This can be calculated by setting the derivative of fitness to zero and then solving for x:

$$\left.\frac{dF(x)}{dx}\right|_{x=x^*} = f(s(1 - x^*) + s_0) - s(fx^* + f_0) = 0, \tag{3.2}$$

from which we have the optimal trait being, $x^* = (fs + fs_0 - f_0s)/(2fs)$.

Another classic optimality model is the optimal foraging theory. Due to the trade-off in foraging time between searching and then consuming resources, the choice of a profitable food item becomes the key to ensure a consumer's survival and thus reproductive success (i.e. fitness). Optimal foraging theory relies on the optimisation of the energy intake rate (*EIR*) subject to time and effort constraints while assuming complete information on the profitability and encounter rate of different resources (Pulliam 1974). For instance, let us assume that a predator has two prey options with densities N_1 and N_2, energy benefit per unit density b_1 and b_2 and handling time h_1 and h_2. Following a Holling type II functional response for the predator, representing a simple partition of the foraging time between the two prey items, we can formulate its total energy intake rate *EIR* as follows:

$$EIR(a) = \frac{ab_1N_1 + (1-a)b_2N_2}{1 + ah_1N_1 + (1-a)h_2N_2}, \tag{3.3}$$

where a represents the probability that the predator will attack the first prey item when encountered. The optimality problem thus reads:

$$\max_{0 \leq a \leq 1} EIR(a), \tag{3.4}$$

which leads to the so-called zero-one rule; that is, the predator will always attack only the prey with higher benefit-to-cost ratio ($b_iN_i/(1 + h_iN_i)$) and ignore the other one, even if encountered. Recent modifications to the classic optimal foraging theory focus on elucidating simple and flexible behavioural strategies that could maximise the *EIR* in changed or changing environment (Zhang and Hui 2014).

3.4   Evolutionary Game Theory

One of the main limitations of optimality models is that fitness only depends on the trait (or the strategy) of the focal individual. Traits and abundances of other interacting individuals are ignored. Models based on game theory have been developed to overcome this limitation. Game theory models consider fitness as a function of not only the strategy of the actor but also the strategies and frequencies of all players in the population. The most well-known case is that of the prisoner's dilemma (PD) for explaining the evolution of cooperation or altruistic behaviours. Two criminals are held in separate cells for the same crime. They can either testify against each other (to defect, D) or remain silent (to cooperate, C; [note, not with the police]). If both remain silent, the two would get only 1-year sentences, due to a lack of evidence (the reward for cooperation). If one stays silent but the other testifies against him, he would be given a major punishment for the crime and sentenced to 10 years in jail, while the defector will be released. If they both testify against each other, both get 3 years in jail. The PD game can be represented by the following pay-off matrix:

$$
\begin{array}{ccc}
\text{Player1\textbackslash 2} & \text{Cooperate} & \text{Defect} \\
\text{Cooperate} & \begin{pmatrix} -1\,(R) & -10\,(S) \\ \text{Defect} & 0\,(T) & -3\,(P) \end{pmatrix}
\end{array}
$$

where R, S, T and P represent the pay-off of player 1 when facing a specific strategy by player 2 (normally known as reward, sucker, temptation and punishment, respectively); PD game assumes $T > R > P > S$. Without knowing the strategy of player

2, player 1 could expect to have an average pay-off of $(R + S)/2 = -5.5$ if being cooperative and an average pay-off of $(T + P)/2 = -1.5$ if being defective. Consequently, being defective is the dominant strategy in a PD game, and the dominant strategy pair is the Nash equilibrium, (D, D). That is, without information of the other, a player better play defectively. However, the Nash equilibrium is not the strategy pair that can yield the minimum total punishment to both players (i.e. maximum total pay-off) which can be achieved when both are being cooperative, (C, C). This often leads to the so-called tragedy of the commons, where even though altruistic/cooperative behaviours can lead to greater societal and eventually personal gain, individual players will still follow the Nash equilibrium and behave selfishly, leading to the depletion of public resources. Note that a defective individual can take advantage of an altruistic society, but an altruistic individual cannot survive in a selfish society. So, how do those notable altruistic behaviours survive in nature?

The game between a hawk and a dove (HD, or Snowdrift) describes the competition for a single food item. Two doves could share the food item (both gaining $R = 1$); a hawk would chase away the dove and obtain the entire food item ($T = 2$ for the hawk and $S = 0$ for the dove) but would fight another hawk for the food item and get injured ($P = -5$). In the HD game, we have $T > R > S > P$, and thus the game has two Nash equilibria: (C, D) and (D, C). That is, the two strategies can coexist, as both strategies have an advantage over the other when rare: A hawk can gain all food items in a dove population without fighting, while a dove will win in a hawk population as all hawks suffer from injuries from each other. This mechanism of having greater pay-offs (fitness) when rare is called negative frequency-dependent selection.

Evolutionary game theory extends the game theory framework above to populations (Maynard Smith and Price 1973). Let us consider the classic formulation that assumes an infinitely large and well-mixed population, where individuals meet in pairs according to their frequencies n_i (law of mass action, Dercole 2016) and then play a single round of a 2-player game, so each receives a certain pay-off. Each strategy increases in frequency if it bears a greater pay-off than the average pay-off to the entire population. Therefore, the expected pay-off for a single individual being cooperative is $P_C = Rn_C + Sn_D$, while the expected pay-off for a defective individual is $P_D = Tn_C + Pn_D$. Since $n_C + n_D = 1$, we only need to consider a single frequency $n_C \equiv n$ ($n_D = 1 - n$). The dynamics of strategy frequency in the population can be captured by the replicator equation:

$$\dot{n} = n(P_C - \bar{P}),\tag{3.5}$$

with $\bar{P}(=nP_C + (1-n)P_D)$ the average pay-off in the population. Therefore, we have $\dot{n} = n(1-n)(n(R-T) + (1-n)(S-P))$. By setting $\dot{n} = 0$, we can identify two boundary strategies $n^* = 0$ and $n^* = 1$, plus an internal strategy:

$$n^* = \frac{S-P}{T-R+S-P}.\tag{3.6}$$

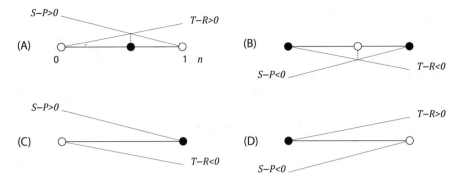

Fig. 3.1 The four possible outcomes of a replicator equation (RE) describing the dynamics of a two-player game. Full (respectively, empty) dots are stable (respectively, unstable) equilibria. (**a**) represents the outcome of selection with the hawk and dove game; (**b**) represents mutual exclusion (priority effect); (**c**) represents dominance of cooperators; (**d**) represents the outcome of the prisoner's dilemma

Consequently, we can distinguish four cases (Fig. 3.1): If $(S - P)(T - R) > 0$, we can have the two strategies' coexistence (case A) when $S > P$ and mutual exclusion (case B) when $S < P$. If $(S - P)(T - R) < 0$, we have the dominance of cooperatives (case C) when $S > P$ and the dominance of defectors (case D) when $S < P$. In the PD game $(T > R > P > S)$, we have $(T - R) > 0$ while $(S - P) < 0$ and thus the dominance of defectors, while in the HD game $(T > R > S > P)$, we have $(S - P) > 0$, thus coexistence.

Several mechanisms have been proposed to explain the evolution of cooperation and summarised as the Hamilton rule, stating that for cooperation to survive, the temptation to defect (T), after discounted by a factor (k), needs to be greater than what a cooperative player will lose to a defector $(-S)$, $kT > -S$. This becomes clearer if we specify the pay-off matrix based on the benefit and cost of the game. Let us assume that a cooperative investor puts c amount of capital into the game while a defector puts zero capital. The return of the capital is b ($>c$). If both players are cooperative, each will have $b - c$ in return. If both players are defective, each gets zero in return (because there is no capital investment). If a defector is playing the game with a cooperative player, the cooperator will lose all the capital, and the defector will gain all the return. Therefore, we have $R = b - c, S = -c, T = b$ and $P = 0$. Since $T > R > P > S$, defection would dominate leading to the tragedy of the commons. The Hamilton Rule states that the necessary condition for cooperation to become the Nash equilibrium is:

$$k \cdot b > c. \qquad (3.7)$$

In kin selection, k is the genetic relatedness between individuals (the classic Hamilton Rule; Hamilton 1964; Maynard Smith 1964). In the theory of direct reciprocity, k is the probability of playing another round of the game between the same two individuals (Axelrod and Hamilton 1981). In the theory of indirect reciprocity, k is the probability of knowing the social reputation of the opponent (Nowak and

Sigmund 1998). In games on graphs or networks, k is the reciprocal of the average number of neighbours (Ohtsuki et al. 2006). In spatial games, k is the proportion of cooperators among the neighbouring individuals of a cooperator (Zhang et al. 2010). Evidently, factors promoting assortative interactions between players can increase the factor of k and thus promote cooperative behaviours in the game. Each of these theories highlights one specific factor in promoting assortative interactions and cooperation: kinship, individual recognition, social tagging, low network degree and cooperation clusters due to harsh environment, respectively.

3.5 Adaptive Dynamics

Adaptive dynamics (AD) is the main framework for modelling phenotypic trait evolution. It extends frequency-dependent selection in evolutionary game theory to *density*-dependent selection and can be considered the evolutionary game theory for continuous traits. AD develops the concept of invasion fitness based on underlying ecological interactions, as well as standard mutation and selection processes. This allows the species to move in a dynamic fitness landscape, where diverse evolutionary dynamics can emerge, such as evolutionary branching, evolutionary suicide and traps (Zhang et al. 2013), as well as Red Queen dynamics (Dercole and Rinaldi 2008). The strength of AD is its ability to elucidate how evolution can push traits to evolve towards either fitness maxima or minima, where the latter could trigger disruptive selection and thus sympatric diversification via evolutionary branching (Geritz et al. 1997, 1998).

The AD models rely on a clear formulation of the invasion fitness – the exponential growth rate of a vanishing mutant in a resident population sitting at its demographic equilibrium (Metz et al. 1992). The fitness function, contrary to optimality models, has to be assessed based on underlying ecological interactions at the time. To do so, we need to (i) specify trait-mediated ecological interactions, (ii) define the resident-mutant competition model and (iii) use the per capita mutant growth rate as the invasion fitness, evaluated when the mutant population size is vanishing (zero) and the abundances of resident populations (or species) are at their equilibria. Evidently, the invasion fitness is a function of the traits of both the mutant and the resident populations, as well as the abundance equilibria of the resident populations (hence the density dependence). Thus, a successful invasion of a mutant trait into a resident population implies resident substitution, whereby the mutant traits substitute and expel the resident population thus becoming the new resident trait; in the fitness landscape, this represents an incremental change of the resident trait and, also importantly, the slight change of the fitness landscape (Fig. 3.2). For this reason the concept of fitness in AD is called adaptive (Metz et al. 1996). The formulation of invasion fitness allows us to study evolutionary trait dynamics using the canonical equation of AD (Dieckmann and Law 1996), essentially describing the selection gradient at the current resident trait value; mathematically, it is the first derivative of the invasion fitness function with respect to the mutant trait, evaluated at the present resident trait value.

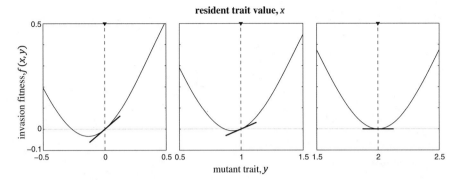

Fig. 3.2 Invasion fitness $f(x, y)$ (Eq. 3.10) as a function of mutant trait, y, for different values of the resident trait x to be invaded (vertical dashed line). The selection gradient is highlighted as the slope of the mutant fitness at the resident trait. Parameter values: $r = K_0 = 1$, $x_0 = 0$, $\sigma = \beta = 0.5$ and $\sigma_K = 2$

The AD of a particular eco-evolutionary system can be numerically solved following this iterative procedure: Once the selection gradient vanishes at a particular trait value, known as evolutionary singularity, the curvature (second derivative) of invasion fitness can tell whether evolution ceases or continues. That is, at a point of evolutionary singularity, a fitness maximum represents the trait to be evolutionarily stable (an evolutionarily stable strategy, ESS), while a fitness minimum represents the trait to be evolutionarily unstable (Della Rossa et al. 2015; Dercole et al. 2016). For an evolutionarily unstable singularity, an evolutionary branching may occur, often requiring additional conditions of protected dimorphism (Dercole and Geritz 2016) and assortative interactions or mating (Dieckmann and Doebeli 1999). At a point of evolutionary branching, a higher dimensional system has to be defined to replace the previous dynamical system (Landi et al. 2013, 2015; Landi and Dercole 2016), normally with an extra equation of a new resident population included; note that simultaneous branching into more than two morphs is theoretically possible but extremely rare. The procedure of AD can then start again for modelling further evolutionary dynamics (Hui et al. 2015, 2017; see also Wilsenach et al. 2017 for an extended formulation of AD).

As an example we use the model of resource competition (Doebeli and Dieckmann 2000; Gallien et al. 2018). Individuals in a population (or community) are competing for resources, with the strength of their competition determined by one continuous scalar trait x (e.g. plant height or body size). Resources are represented by the carrying capacity for different traits, using a Gaussian function along the trait axis, $K(x) = K_0 \exp\left(-(x - x_0)^2 / (2\sigma_k^2)\right)$, representing a single maximum K_0 around an optimal trait x_0 (hereafter assume $K_0 = 1$ and $x_0 = 0$, Fig. 3.3a). Competition for resources is trait-mediated and asymmetric: Individuals with larger trait values are better competitors. However, this increase in competitive ability is often offset by a reduced capacity for resource uptake for $x > x_0$. This generates a competition trade-off, similar to what occurs in nature (e.g. large body

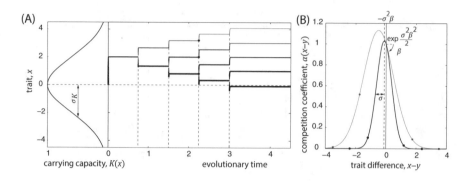

Fig. 3.3 (**a**) Left, carrying capacity function ($K_0 = 1$, $x_0 = 0$, $\sigma_K = 2$); right, trait evolution up to five morphs. Other parameter values: $r = 1$ and $\sigma = \beta = 0.5$. (**b**) Competition function (thick line, $\sigma = 0.5$; normal line, $\sigma = 1$)

size helps for competition but requires more resources to sustain). Competition strength also declines with increasing trait differences between the two involved individuals (larger individuals tend to reduce competition with much smaller ones). This trait-mediated asymmetric competition of phenotype y on x (identical to the competition coefficient α_{xy} in a standard Lotka-Volterra model) is defined by the following (Fig. 3.3b):

$$\alpha(x - y) = \exp\left(\frac{-(x - y)^2}{2\sigma^2}\right) / \exp((x - y)\beta), \tag{3.8}$$

where σ and β describe the breadth and asymmetry of competition, respectively. A greater value of σ extends niche overlap and thus increases the competition coefficient α. A greater value of β gives more advantage to the larger competitor over the smaller one. Note that the intraspecific competition coefficient $\alpha(x - x)$ is 1, meaning that the population size of this phenotypic trait reaches its carrying capacity $K(x)$ in the absence of competition from other traits (Fig. 3.3a).

The population is initially monomorphic and characterised by its trait x. Mutation and selection drive trait evolution, the fate of a mutant being determined by its *invasion fitness*. Resource competition is described by a trait-mediated Lotka-Volterra model; for two competing populations with densities N and M and traits x and y, respectively, we have the following model:

$$\begin{aligned} \frac{dN}{dt} &= rN\left(1 - \frac{N + \alpha(x - y)M}{K(x)}\right) \\ \frac{dM}{dt} &= rM\left(1 - \frac{\alpha(y - x)N + M}{K(y)}\right) \end{aligned}, \tag{3.9}$$

where r is the intrinsic growth rate attained at low densities. Assuming the population M with trait y a mutant population emerging when the resident population N is at

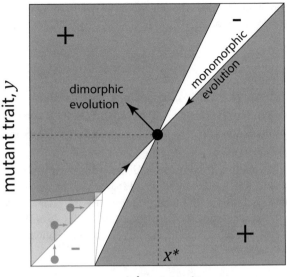

Fig. 3.4 Pairwise invadability plot (PIP). Grey (respectively, white) areas: positive + (respectively, negative (-)) invasion fitness $f(x, y)$. The magnified inset in the lower-left corner represents mutations and resident substitutions when the mutant fitness is positive. This trait substitution sequence is approximated by the monomorphic evolution on the diagonal, eventually converging to the singular strategy x^* (black dot in the centre). Convergence stability of the singular strategy is given by the sign pattern of the fitness below and above the diagonal. In this case, the singular strategy is also a fitness minimum (positive invasion fitness above and below the singular strategy), thus originating dimorphic evolution. Panels in Fig. 3.2 correspond to different vertical sections of the PIP on the left and at the singular strategy

its equilibrium $K(x)$, we can define the mutant invasion fitness as its per capita growth rate at zero mutant density and equilibrium resident density:

$$f(x, y) = \frac{1}{M}\frac{dM}{dt}\bigg|_{M=0, N=K(x)} = r\left(1 - \frac{\alpha(y - x)K(x)}{K(y)}\right). \qquad (3.10)$$

The structure of invasion fitness can be visualised by the zero-contour lines in a pairwise invadability plot (PIP) in the resident-mutant trait space (x, y) (Fig. 3.4). If the invasion fitness of the mutant is higher than the fitness of the resident trait (indicated by the '+' symbol and grey area), the mutant can competitively exclude and replace this resident trait, changing the resident trait from x to y after one incremental step of evolution (in insets of the figure). Trait evolution proceeds with the succession of mutant-resident substitutions along the *selection gradient* (Dieckmann & Law 1996):

$$g(x) = \frac{\partial f(x, y)}{\partial y}\bigg|_{y=x} = r\left(\beta - \frac{x}{\sigma_K^2}\right). \qquad (3.11)$$

The selection gradient describes the directional selection of trait evolution according to the canonical equation of AD:

$$\frac{dx}{dt} = \frac{1}{2}\mu \, s^2 \, \widehat{N}(x) \, g(x),$$

(3.12)

where μ and s^2 are per capita mutation rate and variance of mutational steps (assumed to be normally distributed with mean 0), $\widehat{N}(x)$ is the abundance equilibrium of the resident population $\left(\widehat{N}(x) = K(x)\right)$, and the factor 1/2 expresses that half of the mutations are lost due to negative fitness.

Directional selection will continue until an evolutionary singularity x^* which annihilates the selection gradient $g(x^*) = 0$ is reached; in the case above, we have $x^* = \beta\sigma_K^2$. Note that this evolutionary singularity is greater than the resource optimum $x_0 = 0$ due to assumed competition asymmetry $(\beta > 0)$. The evolutionary dynamics converge to such a singular strategy, since the eigenvalue of the linearised trait dynamics is negative:

$$\frac{d}{dx}\left(\frac{dx}{dt}\right)\bigg|_{x=x^*} = \frac{1}{2}\,\mu\,s^2\,\widehat{N}(x^*)\frac{d}{dx}g(x)\bigg|_{x=x^*} = \frac{1}{2}\,\mu\,s^2\,K(x^*)\left(-\frac{r}{\sigma_K^2}\right) < 0. \quad (3.13)$$

The dynamical stability (attractive) of an evolutionary singularity is known as convergence stability; a strategy that is both convergence stable and evolutionarily stable (fitness maximum) is a continuously stable strategy (CSS; Eshel 1983).

3.6   Evolutionary Branching

An evolutionarily singular strategy x^* represents either a local maximum or minimum in the adaptive fitness landscape. A local fitness maximum signals the end of incremental evolution, whereas a local fitness minimum begets *evolutionary branching*. Evolutionary branching allows the resident and mutant to coexist under protected dimorphism, forming two resident phenotypes diverging under disruptive selection (Fig. 3.5). The latter is known as evolutionary instability (Eshel 1983; Christiansen 1991). The two necessary and sufficient branching conditions at x^* are:

$$\begin{aligned} \frac{\partial^2 f(x^*,y)}{\partial y^2}\bigg|_{y=x^*} &> 0 \\ \frac{\partial^2 f(x,y)}{\partial x \partial y}\bigg|_{y=x=x^*} &< 0 \end{aligned}$$

(3.14)

The first condition implies selection to be disruptive (i.e. evolutionary instability of a fitness minimum), and the second condition implies resident-mutant coexistence

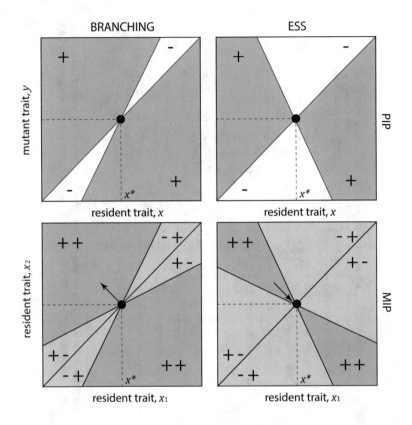

Fig. 3.5 Pairwise (PIP, first row) and mutual invadability plots (MIP, second row), for the two convergence-stable cases of branching (left column) and temporarily invadable ESS (right column; see Table 3.1). MIPs are obtained from PIPs by superimposing their mirror images with respect to the diagonal (i.e. $f(y, x)$). Convergence stability is given by the sign of the invasion fitness around the diagonal of the PIP. Evolutionary stability is given by the sign of the fitness above and below the singular strategy in the PIP. Protection of dimorphism is given by the ++ region centred at the singular strategy in the MIP

(i.e. the protection of dimorphism). In a monomorphic case, the sum of the left side of the two conditions equals the derivative of the selection gradient at the singular strategy,

$$\left. \frac{d}{dx} g(x) \right|_{x=x^*},$$

which is negative if the singular strategy x^* is convergence stable (see Eq. 3.13). That is, convergence stability and evolutionary instability imply the protection of dimorphism in monomorphic populations, thus evolutionary branching. Therefore, we can classify all possible monomorphic singular strategies according to their convergence and evolutionary stability and resident-mutant coexistence (Table 3.1). In a

Table 3.1 Classification of monomorphic singular strategies according to convergence stability, protection of dimorphism and evolutionary stability

Convergence stability		Evolutionary stability	Outcome
$\frac{\partial^2 f(x^*,y)}{\partial y^2}\Big|_{y=x^*} + \frac{\partial^2 f(x,y)}{\partial x\partial y}\Big|_{y=x=x^*} < 0$	Protected dimorphism $\frac{\partial^2 f(x,y)}{\partial x\partial y}\Big|_{y=x=x^*} < 0$	Evolutionary stability $\frac{\partial^2 f(x^*,y)}{\partial y^2}\Big|_{y=x^*} < 0$	Temporarily invadable ESS
		Evolutionary instability $\frac{\partial^2 f(x^*,y)}{\partial y^2}\Big|_{y=x^*} > 0$	Branching point
	Unprotected dimorphism $\frac{\partial^2 f(x,y)}{\partial x\partial y}\Big|_{y=x=x^*} > 0$	Evolutionary stability $\frac{\partial^2 f(x^*,y)}{\partial y^2}\Big|_{y=x^*} < 0$	CSS
Convergence instability $\frac{\partial^2 f(x^*,y)}{\partial y^2}\Big|_{y=x^*} + \frac{\partial^2 f(x,y)}{\partial x\partial y}\Big|_{y=x=x^*} > 0$	Unprotected dimorphism $\frac{\partial^2 f(x,y)}{\partial x\partial y}\Big|_{y=x=x^*} > 0$	Evolutionary stability $\frac{\partial^2 f(x^*,y)}{\partial y^2}\Big|_{y=x^*} < 0$	Convergence-unstable ESS (Garden of Eden)
		Evolutionary instability $\frac{\partial^2 f(x^*,y)}{\partial y^2}\Big|_{y=x^*} > 0$	Convergence unstable evolutionary trap
	Protected dimorphism $\frac{\partial^2 f(x,y)}{\partial x\partial y}\Big|_{y=x=x^*} < 0$	Evolutionary instability $\frac{\partial^2 f(x^*,y)}{\partial y^2}\Big|_{y=x^*} > 0$	Convergence-unstable branching point

ESS evolutionarily stable strategy, *CSS* continuously stable strategy (convergence-stable ESS). At a temporarily invadable ESS, resident and mutant can coexist, but eventually either one of the two will become extinct, or their traits will both converge back at the ESS (see Fig. 3.5). An evolutionary trap has the potential for diversification under disruptive selection, but resident and mutant cannot coexist

polymorphic case, convergence stability is independent from evolutionary stability and dimorphism protection, and thus all combinations are possible (Landi 2014). In the example above, the sole necessary and sufficient condition for evolutionary branching at the monomorphic convergence stable singular strategy x^* is:

$$\frac{\partial^2 f(x^*, y)}{\partial y^2}\bigg|_{y=x^*} = r\left(\frac{1}{\sigma^2} - \frac{1}{\sigma_K^2}\right) > 0. \tag{3.15}$$

This can be simply translated to $\sigma < \sigma_K$; that is, the width of a competition kernel needs to be less than the width of a resource kernel, so that the fitness gain in reducing competition via differentiating traits is greater than the fitness loss from diminishing resource acquisition.

After the first evolutionary branching, we can repeat the same procedure for a higher dimensional system and a similar canonical equation (Landi et al. 2013), although analytic solutions are intractable; we rely on numerical analyses for further inference. In particular, once a dimorphic trait equilibrium is reached, we need to test the branching conditions for both singular traits. If both traits can possibly undergo evolutionary branching, the one with a faster divergence rate (calculated as the product of equilibrium density before branching and the fitness curvature) will in effect undergo an evolutionary branching, while the other will miss the opportunity for dimorphism (Kisdi 1999; Della Rossa et al. 2015). For example, after the first branching (if $\sigma < \sigma_K$), there will be two resident traits with abundances N_1 and N_2 and traits x_1 and x_2 in the system. The resident-mutant model reads:

$$\frac{dN_1}{dt} = rN_1\left(1 - \frac{N_1 + \alpha(x_1 - x_2)N_2 + \alpha(x_1 - y)M}{K(x_1)}\right)$$

$$\frac{dN_2}{dt} = rN_2\left(1 - \frac{\alpha(x_2 - x_1)N_1 + N_2 + \alpha(x_2 - y)M}{K(x_2)}\right). \tag{3.16}$$

$$\frac{dM}{dt} = rM\left(1 - \frac{\alpha(y - x_1)N_1 + \alpha(y - x_2)N_2 + M}{K(y)}\right)$$

The two resident traits coexist at the following equilibrium:

$$N_1^*(x_1, x_2) = \frac{K(x_1) - \alpha(x_1 - x_2)K(x_2)}{1 - \alpha(x_1 - x_2)\alpha(x_2 - x_1)},$$

$$N_2^*(x_1, x_2) = \frac{K(x_2) - \alpha(x_2 - x_1)K(x_1)}{1 - \alpha(x_1 - x_2)\alpha(x_2 - x_1)}, \tag{3.17}$$

and thus the invasion fitness is:

$$f(x_1, x_2, y) = r\left(1 - \frac{\alpha(y - x_1)N_1^*(x_1, x_2) + \alpha(y - x_2)N_2^*(x_1, x_2)}{K(y)}\right). \tag{3.18}$$

For analytical insight, let us assume symmetric competition ($\beta = 0$), which will lead to the two resident traits being symmetric around $x^* = x_0 = 0$. Consequently, we could let $x_1 \equiv -\varepsilon$ and $x_2 \equiv \varepsilon$, and therefore we have $\alpha(x_1 - x_2) = \alpha(x_2 - x_1) = \alpha(2\varepsilon)$ and $K(x_1) = K(x_2) = K(\varepsilon)$. This simplifies the equilibrium $N_1^*(x_1, x_2) = N_2^*(x_1, x_2) = K(\varepsilon)/(1 + \alpha(2\varepsilon)) \equiv \widehat{N}(\varepsilon)$ and invasion fitness $f(\varepsilon, y) = \$\$r\left(1 - \left((\alpha(y + \varepsilon) + \alpha(y - \varepsilon))\widehat{N}(\varepsilon)\right)/K(y)\right)$. Because $\partial\alpha(y - x)/\partial y = -(y - x)\alpha(y - x)/\sigma^2$ and $\partial K(y)/\partial y = -yK(y)/\sigma_K^2$, we can derive the following selection gradient for transformed value ε of trait x_1:

$$g_1(\varepsilon) = \left.\frac{\partial f(\varepsilon, y)}{\partial y}\right|_{y=-\varepsilon} = \frac{-r}{1 + \alpha(2\varepsilon)}\left(\frac{(2\varepsilon)\alpha(2\varepsilon)}{\sigma^2} - \frac{\varepsilon(1 + \alpha(2\varepsilon))}{\sigma_K^2}\right). \quad (3.19)$$

Note that $g_2(\varepsilon) = -g_1(\varepsilon)$, meaning that the selection gradients of the resident traits are opposite in direction and equal in absolute value, thus keeping divergence symmetric around x^*. When selection gradients disappear in the system above, we can compute the evolutionary singularity of transformed trait value:

$$\varepsilon^* = \frac{\sigma}{\sqrt{2}}\sqrt{\ln\left(2\frac{\sigma_K^2}{\sigma^2} - 1\right)}. \quad (3.20)$$

Note that this requires $\sigma < \sigma_K$ which is the condition for the first evolutionary branching in the system (Eq. 3.15). It can be further proved that the dimorphic singularity $(x_1^*, x_2^*) \equiv (-\varepsilon^*, \varepsilon^*)$ is both convergence stable and evolutionarily unstable under the same condition (Birand and Barany 2014). Numerical simulations confirm that there is potentially a cascade of subsequent branching events following the first evolutionary branching in the system (Fig. 3.3a).

3.7 The Price Equation

Recent developments have synthesised a number of theoretical frameworks into the fundamental theorems of evolution based on the Price equation (Queller 2017; Lehtonen 2018). The Price equation can be served as an internode to connect many evolutionary theories, such as Fisher's equation and the Breeder's equation, as well as the AD canonical equation. Let there be individuals with different traits in a population, with $n_{i,t}$ individuals having the trait $x_{i,t}$ at generation (or time) t. Fitness for individuals with trait $x_{i,t}$ can be defined as the ratio between the offspring and parent individuals of this type: $w_{i,t} = n_{i,t+1}/n_{i,t}$. The mean trait value of the population at generation t is $\bar{x}_t = \sum_i x_{i,t}n_{i,t}/\sum_i n_{i,t} \equiv \mathrm{E}(x_{i,t})$, and the mean fitness is $\bar{w}_t = \sum_i w_{i,t}n_{i,t}/\sum_i n_{i,t} = \sum_i n_{i,t+1}/\sum_i n_{i,t} \equiv \mathrm{E}(w_{i,t})$. According to standard statistical probability theory, we can calculate the covariance between the fitness and the trait as follows:

$$\text{cov}(w_{i,t}, x_{i,t}) = E(w_{i,t} x_{i,t}) - E(w_{i,t})E(x_{i,t}). \tag{3.21}$$

We also have the linear operator of expectation:

$$E(w_{i,t} \cdot \Delta_t x_{i,t}) = E(w_{i,t} x_{i,t+1}) - E(w_{i,t} x_{i,t}). \tag{3.22}$$

Combining the equations above, we have $\text{cov}(w_{i,t}, x_{i,t}) + E(w_{i,t} \Delta_t x_{i,t}) = E(w_{i,t} x_{i,t+1}) - E(w_{i,t})E(x_{i,t})$. Note that $E(w_{i,t} x_{i,t+1}) = E(w_{i,t})E(x_{i,t+1})$. Therefore, we have the Price equation:

$$\Delta_t \bar{x}_t = (\text{cov}(w_{i,t}, x_{i,t}) + E(w_{i,t} \Delta_t x_{i,t})) / \bar{w}_t. \tag{3.23}$$

The covariance $\text{cov}(w_{i,t}, x_{i,t})$ captures the direct effect of natural selection: If the trait and the fitness are positively correlated, the mean trait is then expected to increase in the population and vice versa. The second term $E(w_{i,t} \Delta_t x_{i,t})$ represents biased transmission: That is, the fitness advantage of phenotype i is diluted due to imperfect heritability. Without transmission bias and a focus on fitness as the selected phenotype ($x = w$), one could recover Fisher's fundamental theorem of adaptation:

$$\Delta \bar{w} = \frac{1}{\bar{w}} \text{Var}(g(w_i)), \tag{3.24}$$

where g is the individual breeding value, i.e. the proportion of its trait (in this case, fitness w_i) which is passed on to its offspring. In other words, Fischer's fundamental theorem of adaptation states that the change in the average fitness in a population is proportional to the additive variance in fitness. Using Taylor expansion of the fitness and ignoring transmission bias ($\Delta_t x_{i,t} = 0$), we can further connect the Price equation with gradient dynamics, continuous trait game theory and adaptive dynamics, specifically the selection gradient, singular points, convergence and evolutionary stability and the AD canonical equation (Lehtonen 2018). Overall, these sets of models have highlighted the fitness (pay-off) and abundances (frequencies) of different traits as the key variables for depicting evolutionary dynamics.

3.8 Evolutionary Distributions

Evolutionary distributions (ED) study the distribution of phenotypic traits in continuous adaptive spaces. Cohen (2009) pointed out that the mean phenotypic trait in a population may not be adopted by any individuals and thus proposed the use of ED (although the approach was anticipated by Levin and Segel 1985). Furthermore, ED can overcome the timescale separation assumed in different approaches to studying eco-evolutionary dynamics.

ED models commonly take the form of non-local reaction-diffusion equations. The non-local reaction terms are used to incorporate interactions between

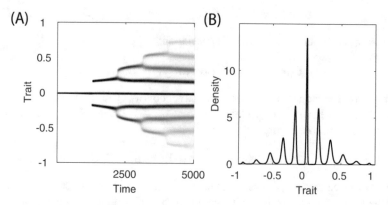

Fig. 3.6 An illustration of evolutionary distribution modelling. (**a**) Trait evolution from an initial population concentrated at $x_0 = 0$. (**b**) Trait distribution at $t = 5000$. Parameter values: $r = 1$, $\sigma = 0.15$, $\sigma_K = 0.5$ and $\mu = 10^{-10}$

individuals with different trait values. Diffusion on the other hand captures trait mutation. The evolution of a species undergoing resource competition can be modelled by:

$$\frac{\partial n}{\partial t} = rn\left(1 - \frac{\int \alpha(x,y)n(y,t)dy}{K(x)}\right) + \mu\frac{\partial^2 n}{\partial x^2}, \tag{3.25}$$

where the competition kernel α denotes the competition strength between individuals of phenotypes x and y; K denotes the resource available to each phenotype, and μ is the trait diffusion rate. Although bimodal resource distribution has been considered in literature, bell-shaped functions are more commonly used for the competition kernel (α) and resource distribution (K) (Sect. 3.5).

Polymorphism occurs in the ED framework when the non-local reaction terms lead to a non-monotonic distribution of the phenotypic traits. A local maximum in the distribution corresponds to a *morph* in the population. Such diversity is sustained by density-dependent selection when the width of the competition kernel (α) is less than that of the resource (K) (Sect. 3.5). In addition, the trait diffusion yields Gaussian-shaped distribution around the local maxima. The emergence of diversity is not limited to the number of morphs but can also include the breadth or the variance of the trait distribution around them. Here the breadth refers to the distance between the local minima surrounding a local maximum (Fig. 3.6).

References

Axelrod R, Hamilton WD (1981) The evolution of cooperation. Science 211:1390–1396
Birand A, Barany E (2014) Evolutionary dynamics through multispecies competition. Theor Ecol 7:367–379

Christiansen FB (1991) On conditions for evolutionary stability for a continuously varying character. Am Nat 138:37–50

Cohen D (1966) Optimizing reproduction in a randomly varying environment. J Theor Biol 12:119–129

Cohen Y (2009) Evolutionary distributions. Evol Ecol Res 11:611–635

Della Rossa F, Dercole F, Landi P (2015) The branching bifurcation of adaptive dynamics. Int J Bifurcat Chaos 25:1540001

Dercole (2016) The ecology of asexual pairwise interactions: the generalized law of mass action. Theor Ecol 9:299–321

Dercole F, Geritz SAH (2016) Unfolding the resident-invader dynamics of similar strategies. J Theor Biol 394:231–254

Dercole F, Rinaldi S (2008) Analysis of evolutionary processes: the adaptive dynamics approach and its applications. Princeton University Press, Princeton

Dercole F, Della Rossa F, Landi P (2016) The transition from evolutionary stability to branching: a catastrophic evolutionary shift. Sci Rep 6:26310

Dieckmann U, Doebeli M (1999) On the origin of species by sympatric speciation. Nature 400:354–357

Dieckmann U, Law R (1996) The dynamical theory of coevolution: a derivation from stochastic ecological processes. J Math Biol 241:370–389

Doebeli M, Dieckmann U (2000) Evolutionary branching and sympatric speciation caused by different types of ecological interactions. Am Nat 156:S77–S101

Eshel I (1983) Evolutionary and continuous stability. J Theor Biol 103:99–111

Gallien L, Landi P, Hui C, Richardson DM (2018) Emergence of weak-intransitive competition through adaptive diversification and ecoevolutionary feedbacks. J Ecol 106:977–889

Geritz SAH, Metz JAJ, Kisdi E, Meszéna G (1997) The dynamics of adaptation and evolutionary branching. Phys Rev Lett 78:2024–2027

Geritz SAH, Kisdi E, Meszéna G, Metz JAJ (1998) Evolutionarily singular strategies and the adaptive growth and branching of the evolutionary tree. Evol Ecol 12:35–57

Gyllenberg M, Parvinen K (2001) Necessary and sufficient conditions for evolutionary suicide. Bull Math Biol 63:981–993

Hamilton WD (1964) The genetical evolution of social behaviour. J Theor Biol 7:1–16

Hui C, Minoarivelo O, Nuwagaba S, Ramanantoanina A (2015) Adaptive diversification in coevolutionary systems. In: Pontarotti P (ed) Evolutionary biology: biodiversification from genotype to phenotype. Springer, Berlin, pp 167–186

Hui C, Minoarivelo O, Landi P (2017) Modelling coevolution in ecological networks with adaptive dynamics. Math Meth Appl Sci. https://doi.org/10.1002/mma.4612

Kisdi E (1999) Evolutionary branching under asymmetric competition. J Theor Biol 197:149–162

Landi P (2014) The emergence of diversity in the adaptive dynamics framework: theory and applications. Politecnico di Milano, PhD Thesis

Landi P, Dercole F (2016) The social diversification of fashion. J Math Sociol 40:185–205

Landi P, Dercole F, Rinaldi S (2013) Branching scenarios in eco-evolutionary prey-predator models. SIAM J Appl Math 73:1634–1658

Landi P, Hui C, Dieckmann U (2015) Fisheries-induced disruptive selection. J Theor Biol 365:204–216

Lehtonen J (2018) The price equation, gradient dynamics, and continuous trait game theory. Am Nat 191:146–153

Levin SA, Segel LA (1985) Pattern generation in space and aspect. SIAM Rev 27:45–67

Maynard Smith J (1964) Group selection and kin selection. Nature 201:1145–1147

Maynard Smith J (1966) Sympatric speciation. Am Nat 100:637–650

Maynard Smith J, Price GR (1973) The logic of animal conflicts. Nature 246:15–18

Metz JAJ, Nisbet RM, Geritz SAH (1992) How should we define fitness for general ecological scenarios? Trends Ecol Evol 7:198–202

Metz JAJ, Gerit SAH, Meszéna G, Jacobs FJA, van Heerwaarden JS (1996) Adaptive dynamics: a geometrical study of the consequences of nearly faithful reproduction. In: van Strien SJ, Verduyn Lunel SM (eds) Stochastic and spatial structures of dynamical systems. Elsevier Science, Amsterdam, pp 183–231

Nowak MA, Sigmund K (1998) Evolution of indirect reciprocity by image scoring. Nature 393:573–577

Ohtsuki H, Hauert C, Lieberman E, Nowak MA (2006) A simple rule for the evolution of cooperation on graphs and social networks. Nature 441:502–505

Parker GA, Begon M (1986) Optimal egg size and clutch size: effects of environment and maternal phenotype. Am Nat 128:573–592

Pulliam HR (1974) On the theory of optimal diets. Am Nat 108:59–74

Queller DC (2017) Fundamental theorems of evolution. Am Nat 289:345–353

Van Valen L (1973) A new evolutionary law. Evol Theory 1:1–30

Wilsenach J, Landi P, Hui C (2017) Evolutionary fields can explain patterns of high-dimensional complexity in ecology. Phys Rev E 95:042401

Zhang F, Hui C (2014) Recent experience-driven behaviour optimizes foraging. Anim Behav 88:13–19

Zhang F, Tao Y, Li ZZ, Hui C (2010) The evolution of cooperation on fragmented landscapes: the spatial Hamilton rule. Evol Ecol Res 12:23–33

Zhang F, Hui C, Pauw A (2013) Adaptive divergence in Darwin's race: how coevolution can generate trait diversity in a pollination system. Evolution 67:548–560

Chapter 4
Networks

Abstract Species are not isolated but interact with other species for survival, forming complex ecological networks, such as food webs and pollination networks. These ecological networks are self-organised in nature for ecosystem functioning, with detectable non-random network structures and architectures. The ubiquity of these structures suggests some universal mechanisms governing the persistence of each species and the functioning of the entire network. Ecological network models have thus been developed to imitate the emergence of the architecture and functioning of real-world networks. While some ecological network models consider the role of long-term coevolutionary dynamics in shaping networks of biotic interactions, others put the emphasis on the adaptability of interaction. Discussions using these models fit the ongoing debate on the relationship between network complexity and stability. This chapter introduces key metrics, models and notions for studying ecological network emergence.

4.1 Graph Theory Basics

A network is a graphic way of representing patterns of interactions between components of a given system. It is composed of a set of N nodes or vertices as the components of the system and L edges or links that connect specific pairs of nodes, the node degree being the total number of edges connected to a particular node (k). Hence, a network is usually represented by either a graph of nodes and edges or by an interaction matrix $\langle a_{ij} \rangle$, also known as an adjacency matrix, its elements representing interaction strengths between nodes i and j. Research on network structure and function is rooted in graph theory, stimulated by the rising theories of random graphs (Erdős and Rényi 1959), which paved the way for the study of non-random properties and the topologies of complex networks.

Studies on the structure of real-world networks have gradually become the mainstay of network literature, where the dominant characteristics of real-world networks have been discussed, authorities tentatively reaching a consensus. For example, networks of friendship, the World Wide Web and metabolic networks of

© The Author(s), under exclusive licence to Springer International Publishing AG, part of Springer Nature 2018
C. Hui et al., *Ecological and Evolutionary Modelling*, SpringerBriefs in Ecology, https://doi.org/10.1007/978-3-319-92150-1_4

cellular biochemicals, all exhibit the *small-world effect*: Most nodes can be reached from any other nodes via only a small number of steps, on average, by only six steps. This phenomenon is commonly known as the principle of 'six degrees of separation' (Milgram 1967). Specifically, in a small-world network, the distance D between two randomly chosen nodes (i.e. the shortest route or the minimum number of steps via existing edges) grows proportionally to the logarithm of the number of nodes N in the network: $D \propto \log N$. The beta model that rewires a regular network with certain amounts of disorder can reproduce the small-world phenomenon (Watts and Strogatz 1998). Moreover, some networks such as the World Wide Web and citation networks also exhibit a scale-free distribution of their node degrees (Barabási and Albert 1999), $P(k) \propto k^{-\gamma}$ especially for large values of k, with the exponent γ often within the range $2 \leq \gamma \leq 4$. Since then many models, such as the preferential attachment model (Barabási and Albert 1999), have been developed to try to explain the mechanisms which lie, potentially, behind the complex structure and dynamics of networks.

In the field of ecology, trophic food webs and networks of predator-prey interactions have long been scrutinised, with the first empirically described food web appearing in literature in the 1920s (Summerhayes and Elton 1923). At the time, Elton used the term *food chain* and named all food chains of a community a *food cycle*. Two other types of ecological networks, those between hosts and parasites and between mutualistic partners, are less well-studied but have nonetheless received great attention in recent years. To date, it has firmly been established that understanding the structure and dynamics of ecological networks is crucial for predicting the fate of ecosystems when facing natural and anthropogenic perturbations.

4.2   Types of Interactions and Networks

An ecological network refers to a collection of biotic interactions between the resident species of an ecosystem. Interaction strength, i.e. the elements a_{ij} of its interaction matrix, represents the effect of species j on species i. Consequently, the type of biotic interaction depends on the relative interaction strengths of the two involved species upon each other. In particular, if $a_{ij} > 0$ and $a_{ji} > 0$, species i and j have positive effects on each other; this interaction is *mutualistic* or *facilitative*. When species i and j have negative impacts on each other (i.e. $a_{ij} < 0$ and $a_{ji} < 0$), the interaction is *competitive*. When one has a negative impact on the other, yet the other has a positive impact on the first (i.e. $a_{ij} > 0$ and $a_{ji} < 0$), the interaction is *antagonistic* or *exploitative*. When one causes harm to the other, while the other stays neutral (i.e. $a_{ij} < 0$ and $a_{ji} = 0$), the interaction is *amensalistic*. When one facilitates the other, while the other stays neutral (i.e. $a_{ij} > 0$ and $a_{ji} = 0$), the interaction is *commensalistic*.

A network is unweighted when interaction strengths are represented by a binary variable (0 or 1) describing the absence or presence of an interaction; now, the adjacency matrix $\langle a_{ij} \rangle$ becomes an incidence matrix, also known as a qualitative matrix. A network is weighted when interaction strengths are measured as real

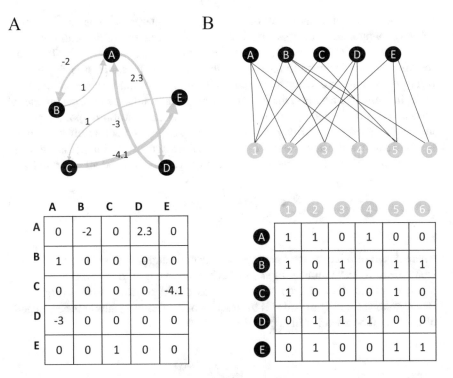

Fig. 4.1 Graphical representation of networks and their adjacency matrices. (**a**) A weighted directed unipartite network such as a food web; (**b**) an unweighted undirected bipartite network, such as a pollination network

numbers, with the adjacency matrix a quantitative one. A network is undirected if species i and j affect each other equally. A network is directed if species i affects species j at a different magnitude from the effect of species j on i. A network is bipartite when the nodes can be classified into two distinct nonoverlapping groups and the interactions (edges) exist only between groups; otherwise, the network is unipartite. For instance, a trophic food web is normally described as a weighted directed unipartite network (Fig. 4.1a), while pollination networks between a group of insect pollinators and a group of flowering plants are an example of a mutualistic bipartite network (note competitions between plants or between pollinators are normally omitted) (Fig. 4.1b).

4.3 Network Topology

For small networks consisting of relatively low numbers of nodes and edges, a simple graphical representation of the network with points and arrows/lines can clearly illustrate the network structure (Fig. 4.1). However, for a large network consisting of tens of thousands of nodes, such visual interpretation is uninformative. However, a number of metrics have been developed exactly for this purpose.

Centrality describes the importance of a node within a network. There are multiple metrics designed to measure node centrality. The simplest is *degree centrality*, or *node degree*. It accounts for the total number of links connected to a focal node. In a directed network, the indegree of a node accounts for the total number of links pointing *to* the node, while the outdegree is the number of links pointing *out* of the node. *Closeness centrality* measures the capacity of a node to affect all other nodes in a network. The closeness C_i of a node i in a network of S nodes is measured as the reciprocal of the average shortest path between node i and all other nodes:

$$C_i = \frac{S-1}{\sum\limits_{j=1, j \neq i}^{S} d_{ij}}, \tag{4.1}$$

where d_{ij} denotes the shortest path between nodes i and j, with the length of a path being measured as a number of steps of consecutive edges. *Betweenness centrality* measures the number of times a node acts as a bridge between two other nodes. Betweenness centrality of node i is the fraction of the shortest paths between all pairs of nodes in the network which pass through node i:

$$B_i = 2 \frac{\sum\limits_{j \neq i \neq k}^{S} g_{jk} g_{jk}(i)}{(S-1)(S-2)}, \tag{4.2}$$

where S is the total number of nodes in the network, g_{jk} is the number of shortest paths between node j and node k and $g_{jk}(i)$ is the number of shortest paths between j and k that pass through node i.

Species richness (S) is the most basic measure of network topology, defined as the total number of nodes in a network, also known as the network size. *Connectance* (C) measures the proportion of realised interactions among all possible interactions: Hence, for a unipartite network, $C = L/(S^2)$, where L is the number of links; note intraspecific interactions are included – whereas, in a bipartite network, $C = L/(m+n)$, m and n are the numbers of species in the two distinct groups, respectively. Sometimes, *connectivity*, or simply the total number of interactions (i.e. L) in a network, is used in place of connectance.

Linkage density (LD) measures the level of generalisation of a network; that is, whether the network is dominated by generalists (species with large node degrees) or specialists (species with small node degrees). What it counts is the average number of links per species: $LD = L/S$. *Weighted linkage density* (LD_w) is the quantitative version of linkage density and is based on the Shannon entropy. In a unipartite network, we have

$$\mathrm{LD_w} = \frac{1}{2}\left(\sum_{k=1}^{S} \frac{a_{k\cdot}}{\bar{a}} 2^{H_{O,k}} + \sum_{k=1}^{S} \frac{a_{\cdot k}}{\bar{a}} 2^{H_{I,k}}\right), \tag{4.3}$$

where $a_{k\cdot}$ and $a_{\cdot k}$ represent the column sum and the row sum of the adjacency matrix $\langle a_{ij} \rangle$ for species k, respectively, $\bar{a}$ is the matrix total, $H_{O,k}$ represents the Shannon diversity index for outflows from species k and $H_{I,k}$ is the Shannon diversity of inflows to species k:

$$H_{O,k} = -\sum_{i=1}^{S} \frac{a_{ik}}{a_{\cdot k}} \ln \frac{a_{ik}}{a_{\cdot k}}$$
$$H_{I,k} = -\sum_{j=1}^{S} \frac{a_{kj}}{a_{k\cdot}} \ln \frac{a_{kj}}{a_{k\cdot}} \tag{4.4}$$

Weighted connectance (C_w) is the quantitative version of connectance $C_w = \mathrm{LD_w}/S$.

The *node degree distribution* ($P(k)$) of ecological networks follows more complicated patterns than the scale-free networks (Barabási and Albert 1999). Most food webs display exponential degree distribution:

$$P(k) \propto e^{-\gamma k}, \tag{4.5}$$

with those of high connectance showing a uniform distribution: $P(k) = 1/(b-a)$ if $a \le k < b$ and $P(k) = 0$ otherwise, where γ, a and b are constant parameters (Camacho et al. 2002; Dunne et al. 2002a). The degree distribution of mutualistic networks mostly follows a decaying exponentially truncated power law with constant parameters γ and α (Jordano et al. 2003):

$$P(k) \propto e^{-\gamma k} k^{-\alpha}. \tag{4.6}$$

Node strength is the quantitative extension of node degree for weighted networks. It is measured as the sum of interaction strengths of all links connected to a node. There are no standard ways of measuring interaction or node strength in ecological networks. For example, the interaction strength in a pollination network may refer to the frequency of pollination visits of a pollinator to a plant. For a food web, it can be measured as biomass flow or consumption rate. The per capita interaction strength is measured as the short-term effect of one individual of a species on one individual of another species.

Clustering coefficient (CC) or *transitivity* measures the extent to which neighbours of a node are also connected. The local clustering coefficient of a node i with k number of neighbours is first defined as $CC_i = L_i/T$, where L_i is the number of links connecting neighbours of i and T the number of possible interactions among neighbours of i. If the network is undirected, the number of possible links between

k nodes is $T = C_k^2 = k(k-1)/2$, i.e. the binomial coefficient of k choose 2. The network clustering coefficient is then computed as the average CC_i over all nodes:

$$CC = \frac{1}{N} \sum_{i=1}^{N} \frac{2L_i}{k(k-1)}. \tag{4.7}$$

Average path length measures the efficiency of information transfer in a network. The shortest path length between two nodes is the minimum number of edges connecting the two nodes. The average path length of a network is thus the average of the shortest path lengths for all pairs of nodes in the network.

4.4   Network Architecture

Modularity or *compartmentalisation* depicts the extent to which interactions are organised into subsets of nodes or modules, where nodes interact more frequently with nodes of the same module than with nodes from other modules. The property of modularity is a common feature of food webs (Moore and Hunt 1988) and pollination networks with high species richness (Olesen et al. 2007). Although a number of metrics have been developed to quantify the level of compartmentalisation in a network, the index of *modularity* (Newman and Girvan 2004) has become the most widely accepted. This measure assumes that nodes in the same module have more links between them than expected for a random network. The *modularity* index is given by:

$$Q = \sum_{i=1}^{m} \left(\frac{l_i}{L} - \left(\frac{d_i}{2L} \right)^2 \right), \tag{4.8}$$

where l_i denotes the number of edges in module i, d_i the sum of the degrees of the nodes in module i and L the total number of edges in the network. However, the limitations of the modularity index are still being debated (Rosvall and Bergstrom 2007; Landi and Piccardi 2014).

Calculating the level of modularity for a network requires us first to divide nodes into a specific module partition and then compute the modularity index (Q) for the particular partition. An optimisation algorithm is normally used for finding the partition that maximises the level of modularity. A number of optimisation algorithms exist for computing network modularity. One of the most commonly used is the *simulated annealing* algorithm (Guimerà and Amaral 2005). It is a stochastic optimisation technique that aims to minimise a given cost function. In the case of modularity, the cost function is $C_t = -Q$. Starting from each node in one module, the algorithm then tries to shift module membership by adding or removing one member node to or from a module and chooses the one that minimises the cost function from a few trials at each step. Note a number of partitions are tried per step

Fig. 4.2 A perfectly nested network. Black cells indicate the presence of interactions, while grey cells the absence of interactions

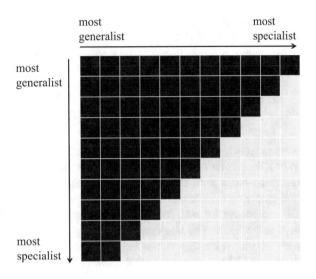

to avoid getting stuck in a local minimum on the cost surface. The final solution is the partition showing a maximum modularity Q_{max}. Among other algorithms, the genetic algorithm and the extremal optimisation algorithm are often used (Schelling and Hui 2015).

Nestedness is a hierarchical pattern of interactions in which specialists can only interact with those species with which generalists interact. In a nested network, both generalists and specialists tend to interact with generalists, whereas specialist-to-specialist interactions are extremely rare (Fig. 4.2). A high level of nestedness is a common feature of mutualistic networks (Bascompte et al. 2003). Several metrics have been developed to quantify nestedness. Among the most commonly used is the *temperature* metric (Atmar and Patterson 1993), with more recent works shifted to using the NODF metric (nestedness metric based on overlap and decreasing fill: Almeida-Neto et al. 2008). The NODF metric takes into account two components: the decreasing fill (DF) of the adjacency matrix and the paired overlap (PO) of its elements. For an *m*-by-*n* interaction matrix of a bipartite network, the nestedness metric is first computed for each pair of rows and columns. Assuming that a pair of rows r_1 and r_2 (respectively, a pair of columns c_1 and c_2) and that r_1 is located above r_2 (respectively, c_1 to the left of c_2), the NODF metric penalises the situation where r_2 is fuller than r_1 (respectively, c_2 is fuller than c_1). Hence, the decreasing fill (DF) metric of the pair of rows (respectively, columns) depends primarily on their marginal totals (MT):

$$\mathrm{DF}_{r_1r_2} = \begin{cases} 100\ (MT_{r_2} > MT_{r_1}) \\ 0\ (MT_{r_2} \leq MT_{r_1}) \end{cases}. \tag{4.9}$$

For non-penalised pairs (i.e. those with $MT_{r_2} > MT_{r_1}$), the paired overlap (PO) metric is computed as the percentage of filled cells in r_2 that are also filled in

r_1. The NODF value for the pair of rows is then computed as $\text{NODF}_{r_1 r_2} = \text{DF}_{r_1 r_2}$ $\text{PO}_{r_1 r_2}$ (respectively, $\text{NODF}_{c_1 c_2} = \text{DF}_{c_1 c_2} \text{PO}_{c_1 c_2}$). The NODF of the entire matrix of m rows and n columns is then given by the average NODF values of all possible pairs of columns and rows:

$$\text{NODF} = \frac{\sum_{xy} \text{NODF}_{xy}}{\frac{m(m-1)}{2} + \frac{n(n-1)}{2}}, \tag{4.10}$$

where xy corresponds to each pair of columns or rows. A weighted NODF, called WNODF, can be used for estimating the level of nestedness of weighted networks (Almeida-Neto and Ulrich 2011). To calculate the WNODF, the number of non-empty cells is considered, rather than the marginal totals, when computing the DF component for each pair of rows and columns, while the computed PO component is the fraction of non-empty cells in r_2 (respectively, c_2) that have values less than the corresponding values in r_1 (respectively, c_1). NODF and WNODF range from 0 to 100 and increase with the level of nestedness.

4.5   Structural Emergence Models

To comprehend observed network structures, various structural emergence models have been proposed – these models mimic certain key structuring processes: For food webs, there are two standard stochastic models, the cascade model (Cohen et al. 1990) and the niche model (Williams and Martinez 2000). Considering only species richness (S) and connectance (C) as input parameters, the cascade model is based on two rules: (1) Species are placed in a one-dimensional feeding hierarchy; (2) species can only feed on those that are lower in the hierarchy than themselves. The second rule means that lower-left elements of an adjacency matrix, including the diagonal, are set to zero (i.e. $a_{ij} = 0$ for $i \geq j$). The rest of the elements in the matrix can be randomly assigned to 1 with probability $2SC/(S - 1)$ and 0 otherwise. The cascade model performed reasonably well when predicting food web topology, showing that the observed non-random structures could emerge from simple stochastic rules. The niche model relaxes the second rule of the cascade model by allowing a species to feed only on those within its feeding range, rather than any of those lower in the feeding hierarchy. In particular, for species i at a randomly drawn feeding hierarchy n_i, the feeding range r_i is assigned by using a beta function to draw values at random from the interval $(0, 1)$ whose expected value is $2C$ and then multiplying that value by n_i to obtain a web with C that matches the desired connectance. The centre of the feeding range c_i is uniformly drawn from the interval $[r_i/2, n_i]$ or from the interval $[r_i/2, 1 - (r_i/2)]$ if $n_i > 1 - r_i/2$.

By relaxing the one-dimensional feeding hierarchy of the cascade and the niche models, the nested-hierarchy model establishes links based on phylogenetic constraints and allows a link to occur between species that are greatly different in their niche positions (Cattin et al. 2004). In particular, the nested-hierarchy model follows the following pseudocode:

- Assign each species a random niche value from 0 to 1.
- For species i, first assign a random prey j ($j < i$).
- For species i, assign an additional prey k ($k < i$) if the prey k does not have consumers.
- If the additional prey k has other consumers, then select a prey at random from the diet of such consumers that also consume j, and replace prey k by this selected prey.
- If the additional prey k has other consumers but none prey upon j, choose k randomly from $k < i$, otherwise, from $k > i$.
- Repeat the three steps above until the number of prey objects of species i follows a beta distribution that produces a network of connectance C.

However, besides the complexity, its improvement on predicting the structure of natural food webs is not substantial. Evidently, these structuring processes in such models are topological, different from typical ecological processes essential to community or network assembly, such as biotic interactions and evolution.

Biotic interactions in ecological networks are dynamic and adaptive, changing and fine-tuning along evolutionary time scales, during which the richness of network assemblages also fluctuates with the birth and death of species. As such, the evolutionary history of ecological networks could dictate network architecture. In particular, phylogenetic history could reveal how species interacted with each other and how networks were structured deep in the past. As biotic interactions are often closely linked to morphology and thus underlying genotypes of involved species, they could potentially persist over evolutionary time scales, and newly emerged species through coevolution could inherit those interaction partners of their ancestors (Rezende et al. 2007). This means that we could model biotic interactions in long-standing ecological networks as a continuous-time Markov process, along the corresponding phylogenies of the network species assemblage (Minoarivelo et al. 2014).

For a bipartite network, instead of considering the phylogeny of each group of species separately, we could combine their phylogenies and consider whether two species (from each group) interact as a binary trait; by doing so, the network model becomes essentially similar to any phylogenetic modelling of trait evolution (Fig. 4.3; Minoarivelo et al. 2014). In particular, the Markov process model explores a set of two states: S_0 refers to the absence of an interaction and S_1 the presence of an interaction. The state S_j of an interaction at a particular node n_2 of the joint phylogeny is inherited from the state S_i of its parent node n_1. The inheritance is not perfect and depends on the transition probability, $P_{ij}(t) = \exp(q_{ij}t)$, where t is given by the branch length from node n_1 to node n_2; q_{ij} represents the instantaneous rate:

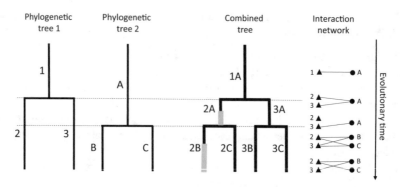

Fig. 4.3 An example of the combination of two phylogenetic trees into a joint tree of interaction. The right panel shows the resulting interaction networks. (Modified from Minoarivelo et al. (2014))

$$\langle q_{ij} \rangle = \begin{array}{c} \\ S_0 \\ S_1 \end{array} \begin{array}{c} S_0 \quad S_1 \\ \left[\begin{array}{cc} -\mu\pi_1 & \mu\pi_0 \\ \mu\pi_0 & -\mu\pi_1 \end{array} \right], \end{array} \qquad (4.11)$$

where μ is the gain-loss rate and π_0 and π_1 are the equilibrium frequencies of S_0 and S_1, respectively. With μ, π_0 and π_1 inferred from observed phylogenies and the interaction matrix, this model could explain 34% variation in the topology and architecture of mutualistic networks. Although this signals the structuring role of phylogenies in ecological networks, we could speculate that there might be important processes missing in such coevolutionary models – which might explain the remaining majority of network variation.

4.6   Network Stability

The function, especially the stability, of an ecological network can be evaluated by monitoring how its dynamical behaviour responds to perturbations. The concept of stability is multifaceted. Two main approaches exist to categorise the stability of an ecological network: demographical and topological (Bascompte and Jordano 2006).

The definition of demographic stability is based on the effects of small fluctuations, caused by small perturbations, on a network when it sits close to a system steady state. When arbitrarily small perturbations decay at the proximity of an equilibrium point (the steady state of a system), the system is said to be *locally asymptotically stable* (May 1973). For a nonlinear system of population dynamics, $\dot{N} = f(N)$, let $\bar{N}$ be the population density at equilibrium (i.e. $f(\bar{N}) = 0$). When a small perturbation $(x(t))$ is introduced at the equilibrium, the population density becomes $N(t) = \bar{N} + x(t)$, and the dynamics of this perturbation can be given by $\dot{x} = \dot{N} = f(N)$. Using linearisation at $\bar{N}$, the dynamics of the perturbation can be approximated by $\dot{x} = Jx(t)$, with J representing the Jacobian matrix of the population dynamics evaluated at the equilibrium $\bar{N}$. The local asymptotic stability of the system depends on the sign of the real part of the leading eigenvalue $\text{Re}(\lambda)$ of the Jacobian: If $\text{Re}(\lambda) > 0$, the perturbation is amplified, and the system is unstable. If $\text{Re}(\lambda) < 0$, the perturbation dies out, and the system returns to its equilibrium (i.e. locally asymptotically stable). Hence, the condition for a system to be locally asymptotically stable is that $\text{Re}(\lambda) < 0$. For an ecological network, the Jacobian matrix J is equivalent to the so-called community matrix A, with its elements a_{ij} describing the change in the population growth rate of species i, caused by a small perturbation in the abundance of species j. Note, in many ecological studies, the adjacency matrix of interaction strength is taken as a surrogate for the community matrix.

A few metrics have been proposed to measure demographic stability quantitatively. *Resilience* is a measure of how fast the system returns to its equilibrium. The resilience of a system is measured by the absolute value of the leading eigenvalue of the community matrix or more often its logarithmic value: $\text{Res} = \ln|\text{Re}(\lambda)|$. Note that a system needs to be locally asymptotically stable before calculating its resilience. *Persistence* is given by the proportion of coexisting species at equilibrium, relative to the initial number of species: $P = S_{\text{final}}/S_{\text{initial}}$, and is less than one if some species have become extinct before the system reaches its equilibrium. *Reactivity* measures whether a perturbation, moving away from equilibrium, can still initially grow and then be amplified in spite of the system's eventual return to equilibrium. It is calculated as the maximum instantaneous rate at which a perturbation runs away from an equilibrium (Chen and Cohen 2001):

$$R = \max_{\|x\|\neq 0}\left[\left(\frac{1}{\|x\|}\frac{d\|x\|}{dt}\right)\Big|_{t=0}\right], \tag{4.12}$$

where x is the vector of perturbation and the norm $\|\cdot\|$ is the Euclidean distance $\left(\sum_{i=1}^{S} x_i^2\right)^{1/2}$ for a network with S number of species. In addition, when the type of a system's dynamical behaviour (e.g. attractor, limit cycle or deterministic chaos) is not affected by small perturbations of its parameter values, the system will not experience bifurcations and thus is called *structurally stable* (Solé and Valls 1992; Rohr et al. 2014).

Topological stability examines the effects of species removal and addition on network structure and topology. *Robustness* accounts for the ability of a network to defy secondary extinctions from the removal of some species, also known as *deletion stability* (Pimm 1979; Paine 1980). In practice, it can be measured as the fraction or number of species to be removed that can result in 50% species loss in a network (Dunne et al. 2002b). Species removal can be random or targeted (e.g. removing first species with lower or higher node degrees or from lower or higher trophic levels) or considered different fractions of species loss (e.g. Nuwagaba et al. 2017). *Invasibility* describes the response of a network to the invasion of new species (Hui and Richardson 2017). A highly invadable network is more absorbable and susceptible to the introduction of new species. In practice, invasibility can be measured as the amount of opportunity niches in the trait space that allow for positive per capita population growth of rare aliens (Hui et al. 2016; see related concept of invasion fitness in Chap. 3).

4.7 Complexity-Stability Relationship

Ecological intuition tells us that complex communities and networks are more stable than simple ones; that is, network complexity enhances stability. MacArthur (1955) argues that 'a large number of paths through each species is necessary to reduce the effects of overpopulation of one species' and thus that stability increases with the node degree of species. Based on his observations of natural communities, Elton (1958) argues that 'simple communities were more easily upset than richer ones', that is, more subject to destructive oscillations in populations and more vulnerable to invasions. Such ecological intuition of a positive relationship between system complexity and stability was challenged mathematically, especially in the seminal work of May (1972, 1973).

In particular, May depicted community dynamics using first-order differential equations and explored its local asymptotic stability by examining the community matrix $\langle a_{ij} \rangle$. The interaction strength in the matrix was randomly assigned from a normal distribution with zero mean and standard deviation α (named the average interaction strength). The diagonals of the matrix, indicating intraspecific interactions, were set to -1 for self-regulation. In addition, interaction strengths were assigned zeros with a probability $1 - C$. Community complexity was measured by three factors: species richness (S), the connectance (C) of the community matrix and the average interaction strength (α). The community tends to transition sharply from stable to unstable behaviours as the complexity of the system increases, with the ecological community near its equilibrium being almost certain of stability if

$$\alpha(SC)^{1/2} < 1, \tag{4.13}$$

which is known as May's stability criterion. Therefore, it is extremely improbable to observe persistent ecosystems that are species rich (large S) or highly connected (large C); this is directly contrary to ecological intuition.

Allesina and Tang (2012) further fine-tuned May's stability criterion by differentiating between interaction types. In particular, interaction strengths were drawn at random from a distribution with mean E(x) and standard deviation σ. Remarkable differences were found for different types of interactions. Let E($|x|$)/$\sigma = \rho$. For a predator-prey network, of which interaction strengths were randomly drawn from a normal distribution with mean E(x) = 0 and with the diagonals of the interaction matrix set to $-d$, the stability criterion is:

$$\sigma(SC)^{1/2} < \frac{d}{1 - \rho^2}. \tag{4.14}$$

For a network composed of all mutualistic interactions, the stability criterion is given by:

$$\sigma(S - 1)C < d/\rho. \tag{4.15}$$

This shows that nested mutualistic networks are less likely to be stable than unstructured networks. For a network composed of only competitive interactions, the stability criterion is given by:

$$\sigma(SC)^{1/2}\frac{1 + (1 - 2C)\rho^2}{(1 - C\rho^2)^{1/2}} < d - \sigma\rho C. \tag{4.16}$$

The relationship between network complexity and stability remains a particular focus in ecology research. To resolve the complexity-stability debate, many have developed more realistic models that include more plausible network structures of higher-order nonlinear (McCann et al. 1998; Krause et al. 2003; Okuyama and Holland 2008) and non-random (Neutel et al. 2002; Suweis et al. 2015; van Altena et al. 2016) interactions, further unfolding the conceptual layers of network stability and complexity (Landi et al. 2018).

4.8   Networks of Coevolving Traits

Species can be characterised by phenotypic or morphological traits. Such traits can determine the state of a species' interaction. For example, body size is a determinant trait for the interaction in food webs. For pollination syndromes, a pollinator trait could be its proboscis length, and the floral trait could be the length of the pollen tube. Such traits undergo mutation and evolve through time, affecting the state of interaction and structure of the entire network. In a setting where species are distinguished only by such traits (i.e. morphospecies) (see Chap. 3), child species are produced as mutations of parent species. The adaptive dynamics approach (see Sect. 3.5) is a useful tool for modelling the coevolution of traits. We here give an example of its use to generate a mutualistic network of animal and plant species.

The per capita growth rate of a morphospecies of animal A_i with trait value x_i and a morphospecies of plant P_j with trait value y_j is described by the Lotka-Volterra model for mutualism with a type II functional response, as detailed in Minoarivelo and Hui (2016a):

$$\frac{dA_i}{A_i dt} = f_A(x_i) = r_A - \frac{r_A \sum_k \gamma(x_i, x_k) A_k}{K_A(x_i)} + \frac{\sum_j b_{A_i P_j} w_{A_i P_j} P_j}{1 + h \sum_j w_{A_i P_j} P_j}$$

$$\frac{dP_j}{P_j dt} = f_P(y_i) = r_P - \frac{r_P \sum_k \gamma(y_j, y_k) P_k}{K_P(y_j)} + \frac{\sum_j b_{P_j A_i} w_{P_j A_i} A_i}{1 + h \sum_i w_{P_j A_i} A_i}, \tag{4.17}$$

where r_A and r_P are intrinsic growth rates of animals and plants, respectively, and h the handling time that animals spend for visiting a plant and digesting the nutrients extracted from the plant.

The carrying capacities (K_A and K_P) are Gaussian functions of the traits following Doebeli and Dieckmann (2000): $K_A(x_i) = k_A N\left(x_A^{\max}, \sigma_A, x_i\right)$ and $K_P(y_j) = k_P N\left(y_P^{\max}, \sigma_P, y_j\right)$. The intraspecific resource competition function $\gamma(x_1, x_2) = \exp\left(-(x_1 - x_2)^2/(2\sigma^2)\right)$ is also a Gaussian function and assumes that morphs with similar traits undergo stronger competition. The mutualistic benefit, b_{AP}, reflects the assumption that matching traits bring to each other high profit and is also assumed to follow a Gaussian function of trait difference: $b_{AP}(x_i, y_j) = c \cdot \exp\left(-(x_i - y_j)^2/(2\sigma_m^2)\right)$. The probability of interaction after encounter (w_{Ap} and w_{PA}) depends on both the benefit and the abundance of the involved morphs. For more details on these functions, see Minoarivelo and Hui (2016a).

A resident species can undergo mutations. The invasion fitness of the mutants with traits x_i' and y_j' can be defined by their per capita growth rate when their initial densities are negligible, i.e. $f_A(x_i')$ and $f_P(y_j')$, respectively. The direction of trait evolution is described by the selection gradients: $g_{A_i} = \partial f_A(x_i')/\partial x_i'\big|_{x_i'=x_i}$ and $g_{P_j} = \partial f_P(y_j')/\partial y_j'\big|_{y_j'=y_j}$, and the dynamics of the trait is given by the canonical equation of adaptive dynamics:

$$\dot{x}_i = m_A \tilde{A}_i g_{A_i}$$
$$\dot{y}_j = m_P \tilde{P}_j g_{P_j}. \tag{4.18}$$

From an initial pair of interacting morphospecies of animal and plant, the mutant can become a new morphospecies entering the system if conditions for evolutionary branching (see Sect. 3.6) are satisfied. Subsequent branchings giving rise to new morphospecies can then occur until the system expands into a complex network of several morphospecies. Interaction networks are depicted through time as a quantitative interaction matrix, of which elements (q_{ij}) are defined as the nonlinear functional response, for both the animal i and the plant j (Berlow et al. 2004):

$$q_{ij} = \frac{1}{2}\left(\frac{A_i b_{A_i P_j} w_{A_i P_j} P_j}{1 + hw_{A_i P_j} P_j} + \frac{P_j b_{P_j A_i} w_{P_j A_i} A_i}{1 + hw_{P_j A_i} A_i}\right). \tag{4.19}$$

Fig. 4.4 A mutualistic
network emerging from
coevolving traits of animals
and plants. Parameter
values: $r_A = r_P = 1, h = 0.1,$
$c = 0.1, \sigma_A = \sigma_P = e^{0.75},$
$\sigma_C = e^{-3}, \sigma = e^{-2.25}.$
(Modified from Minoarivelo
and Hui (2016b))

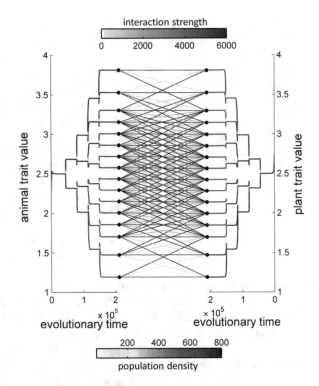

An example of a resulting network that has emerged from trait evolution is given in
Fig. 4.4. Subject to specific conditions on the use of mutualistic resources, such a
model could reproduce some network topology such as high nestedness and
modularity.

4.9 Adaptive Interaction Switching

A species can rapidly change its mutualistic and antagonistic partners in a community, in response to a shifting ecological and environmental context. Evidence for
such *adaptive interaction switching*, or *adaptive rewiring*, is mounting especially in
the field of adaptive foraging, where a species optimises its diet for maximum rate of
energy intake (see Chap. 3; Fossette et al. 2012; Zhang and Hui 2014), and in the
field of interaction fidelity and promiscuity (Sanchez 2015). Different forms of
switching can be found in the available literature: survival through selecting and
switching partners (Staniczenko et al. 2010), decision-making based on profitability
and encounter rate (Zhang and Hui 2014), Wallace's elimination of the unfit (Zhang
et al. 2011; Nuwagaba et al. 2015), and Darwin's survival of the fittest (Kondoh
2003; Suweis et al. 2013). We elaborate here on the last two adaptive interaction
switching models.

Biotic interactions are edges in a network that connect two involved species (nodes). Models of adaptive interaction switching allow the species in a network to change partners (interaction rewiring) based on different adaptive algorithms. According to *the elimination of the unfit*, initially proposed by Alfred Russel Wallace as the process of natural selection, a species suppresses unbeneficial interactions by replacing them with new partners chosen at random. This algorithm of adaptive interaction switching has been implemented for bipartite mutualistic networks (Zhang et al. 2011) and bipartite antagonistic networks (Nuwagaba et al. 2015). For instance, the dynamics of a resource-consumer network can be formulated using a Lotka-Volterra model with a type II functional response:

$$
\begin{aligned}
\frac{1}{R_i}\frac{dR_i}{dt} &= r_i - c_iR_i - \sum_j \frac{a_{ij}\gamma_{ij}N_j}{1 + h\sum_k a_{kj}\gamma_{kj}R_k} \\
\frac{1}{N_i}\frac{dN_i}{dt} &= -d_i + \sum_i \frac{b_{ji}a_{ij}\gamma_{ij}R_i}{1 + h\sum_k a_{kj}\gamma_{kj}R_k}
\end{aligned}
\tag{4.20}
$$

where R_i and N_i are the population density of resource i and consumer j, respectively, r_i and c_i are the intrinsic growth rate and the density-dependent coefficient of resource i, d_i is the mortality rate of consumer j, $\langle a_{ij} \rangle$ is the binary interaction matrix, $\langle \gamma_{ij} \rangle$ is the preference matrix with the element describing the probability of interaction after encounter (note $a_{ij}\gamma_{ij}$ gives the attack rate of consumer j on resource i), $\langle b_{ji} \rangle$ is a matrix of the benefit provided to consumer j from consuming an individual of resource i and h is the consumer's handling time.

Population dynamics is numerically solved, such that, at each time step, a consumer selected at random stops interacting with the resource that contributes the least to its fitness (per capita population growth rate), i.e. the one with the minimum $b_{ji}a_{ij}\gamma_{ij}R_i$. The consumer then starts to exploit another random noninteracting resource. With all parameters initially assigned at random, this model takes only network size and connectance as input. Predictions of this model on modularity, nestedness and node degree distributions fit extremely well with observed values of empirical networks, explaining an amount of 80% network variation (Nuwagaba et al. 2015).

Adaptive interaction switching can of course also be implemented based on the algorithm of *the survival of the fittest*, initially proposed by Herbert Spencer and Charles Darwin for the process of natural selection. A species switches a partner to a new random partner so long as it increases fitness gain – and there is a model for mutualistic networks with additional within-trophic competition (Suweis et al. 2013). At each time step, a species (e.g. animal species i) is selected at random with one of its interactions (e.g. with plant species j). An attempt to switch this interaction to another species (e.g. with another plant species k) is performed. The switch is accepted if and only if the new interaction (i.e. between species i and species k) does not lead to a decrease in the abundance of species i at equilibrium. The population dynamics is also described using Lotka-Volterra models with the Holling functional response. Note each potential switch will need to run the model to a new equilibrium for abundance comparison, which can be time consuming. Modification can be made by replacing the switching condition, from abundance increasing at equilibrium to an increase in instantaneous population growth. Adaptive interaction switching with the survival of the fittest can also explain nested patterns in mutualistic networks, suggesting abundance increments could be driving network emergence (Suweis et al. 2013).

The algorithm implementing the survival of the fittest often leads to extreme network patterns, e.g. predicted nestedness eventually becomes much higher than those observed values, while predicted network patterns from the elimination of the unfit often converge to a stable level, close to observed values (Zhang et al. 2011; Nuwagaba et al. 2017). Note the elimination of the unfit does not guarantee that an interaction switch always warrants instantaneous fitness gain, as the random replacement of the worst partner could be even worse (risk of trying); in contrast, each switching in the survival of the fittest requires an eventual abundance increment. This contrast has led to discussions on two issues. First, the correct process of natural selection could well be *the elimination of the unfit*, instead of the most widely accepted view, *the survival of the fittest*. In addition, a consumer could easily identify the least beneficial resource among its diet, but find it rather difficult or even impossible to assess whether an interaction switch could lead to its abundance gain when the network is settled at equilibrium. This assessment relies on the responses of other species and often incurs extremely long waiting time (to return to the new system equilibrium). Second, ecological networks might not be organised for maximising productivity but resembling a multiplayer game that does not necessarily lead to greater society gains but instead is organised to reduce conflicts between players.

4.10 Meta-Networks

Most studies consider an ecological network of local biotic interactions as a closed system. This is contrary to the reality that local communities and networks are interdependent and are tightly linked by the dispersal of individuals. The concept of the meta-network is rooted in the concept of metapopulation (for regional-level persistence) and meta-community (for regional-level coexistence). A meta-community is a set of local communities connected via the dispersal of individuals from other local communities (Leibold and Chase 2018) or from a regional species pool (Hubbell 2001) of multiple potentially interacting species. By analogy, a meta-network is a set of spatially distributed local networks connected by dispersal and thus influenced by recolonisation and extinction events (Hagen et al. 2012).

We could formulate a meta-network of S species and n local networks by adding dispersal to standard Lotka-Volterra models (Gravel et al. 2016):

A   B

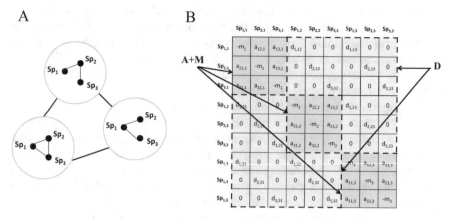

Fig. 4.5 Conceptual example of a meta-network composed of three species at three locations. (a) Its graph representation; (**b**) its interaction matrix which is a superposition of the matrix of interspecific interactions **A**, the matrix of intraspecific interactions **M** and the matrix of dispersal coefficients **D**. (Modified from Gravel et al. 2016)

$$\dot{N}_{ix} = N_{ix}\left(r_{ix} + \sum_{j=1}^{S} a_{ijx} N_{jx}\right) + \sum_{y=1}^{n} d_{ixy}\left(N_{iy} - N_{ix}\right), \qquad (4.21)$$

where N_{ix} and r_{ix} represent the population size and intrinsic growth rate of species i in local network x; a_{ijx} is the per capita effect of species j on species i in local network x and can be randomly drawn, with probability C, from a distribution with a zero mean and standard deviation σ; and d_{ixy} is the dispersal rate of species i from network y to x. Note the Jacobian matrix of the meta-network can be partitioned into three components:

$$\mathbf{J} = \mathbf{M} + \mathbf{D} + \mathbf{A}, \qquad (4.22)$$

where **M** is a diagonal matrix of intraspecific density dependence of $-m_{ix}$, **D** is the dispersal matrix of d_{ixy} and **A** is the collection of local interaction matrices of a_{ijx} with each network a block (see Fig. 4.5 for a meta-network of three species and three local networks). Note C is not the connectance of matrix **A** but only the proportion of nonzero elements for these blocks. With the same dispersal rate and intraspecific density dependence coefficient for all species and interaction strengths between any two species a_{ijx} correlated across networks x with ρ, the correlation coefficient ($\rho = 0$ for high heterogeneity of interactions across networks; $\rho = 1$ for identical interaction strengths across networks), a simplified stability criterion can be derived (Gravel et al. 2016):

$$\sigma(C(S-1)/n_e)^{1/2} < m, \qquad (4.23)$$

with $n_e = n/(1 + (n - 1)\rho)$ the effective number of ecologically independent local networks. Note this is equivalent to May's stability criterion if $\rho = 1$: The maximal admissible complexity, $\sigma^2 C(S - 1)$, is the same as one local network. If $\rho = 0$, meaning the interaction strength of two species is completely context-dependent, the meta-network can allow the maximal admissible complexity to be n times that of a local network. However, when the dispersal rate is low (small d) in a large meta-network (large S and n), the stability criterion above becomes:

$$\sigma(C(S - 1))^{1/2} < m + d. \tag{4.24}$$

That is, weak dispersal can stabilise meta-networks. Overall, a meta-network becomes more stable when interaction strengths are network-specific, or when the dispersal rate is low in large meta-networks. Further analyses could also reveal the role of adaptive interaction switching in local networks, enhancing stability by reducing the standard deviation of strengths of any realised interactions. Moreover, interaction strength a_{ijx} can be trait-mediated, i.e. a function of the trait values of species i and j in network x, which will further allow us to examine trait (co) evolution in meta-networks. These remain ongoing focuses in ecology research.

4.11 Complex Adaptive Networks

A complex adaptive system (CAS) is a dynamical system comprising multiple interacting parts that can adaptively and collectively respond to perturbations. The boundary between the system and its environment is often hard to define as the system and its environment co-adapt and co-evolve. Complexity in system patterns and structures arises from the interrelationship, interaction and interconnectivity of elements within a system and also between the system and its environment (Gell-Mann 1995). As such, CASs are often used for exploring complex behaviours across a wide spatiotemporal scale, emerging from nonlinear abiotic and biotic interactions among the multitude of elements within a loosely bounded open system.

The framework of CAS thus provides a solid foundation for explaining the architecture and function emergence of adaptive ecological networks. Adaptive interaction switching can occur at three different levels. At the individual level, it is important to choose adaptively with whom to interact (habitat and diet selection) or to avoid (anti-predation strategies) (Zhang and Hui 2014). Consequently, inter-action switching can occur rapidly for individuals at a pace even faster than the typical ecological timescale (e.g. the host switch in parasites; van Baalen et al. 2001). At the population level, successful interactions often require involved species to possess a certain level of matching or complementary traits, such as matching phenology for flowering time in plants and the foraging activity of insect pollinators (Waser 2015). Through adaptively changing the level of trait complementarity and plasticity, populations can adjust interaction strength to their own advantage (Santamaría and Rodríguez-Gironés 2007). At the species level, species can completely lose or regain the possibility of interactions by developing incompatible traits of forbidden links and setting up barriers to exploitation (Jordano et al. 2003), potentially through adaptive diversification and evolutionary branching (Dieckmann and Doebeli 1999; Zhang et al. 2013) and geographical barriers. Therefore, processes of interaction switching, together with constraints on network size and complexity (e.g. assemblage history and the coevolution of trait complementarity), are dominant forces that give rise to realised network architecture.

The interaction matrix in recent models of adaptive interaction switching changes constantly, reflecting an adaptive network responding to changes (Staniczenko et al. 2010; Kondoh 2003). By allowing consumers to readjust their exploited resources by updating the interaction matrix, adaptive network models can successfully capture the spontaneous emergence of all features of network architecture. The adaptive interaction switch allows the abundance of species to fluctuate without necessarily leading to extinction when facing perturbations (van Baalen et al. 2001); this is vital for rebalancing the network and returning it to its equilibrium, thereby buffering ecosystems against perturbations (Staniczenko et al. 2010). It is thus a strong structuralising force that largely organises ecological networks at marginal stability, analogy to the system evolution towards the edge of chaos. This makes an adaptive network a better proxy than a rigid network for depicting real ecological communities, and its adaptive nature is important for forecasting the response of ecosystems to environmental changes and perturbations (Valdovinos et al. 2010; Kaiser-Bunbury et al. 2010).

The long-standing debate on the complexity-stability relationship in complex ecological systems can be systematically addressed in various ways (McCann 2000): considering random matrices versus empirical matrices, binary versus weighted matrices or different types of interactions. However, such comparisons treat different features of ecological networks as separate entities, thus ignoring that these features are multifaceted descriptions of network functioning. To appreciate the inherent multifaceticity in depicting network functioning, a potential solution to this debate could be to agree that both architecture and stability features emerge through the spontaneous self-organisation of adaptive ecological networks, a critical feature of CASs. Therefore, the complexity-stability relationship of an ecological network could simply be a by-product of system evolution and functioning.

An ecosystem functions as a grand multiplayer game. To survive in such a game, species must often implement multiple contingency plans to handle ecological or evolutionary selection pressures. Ecologically, species can adjust the extent and structure of their geographical range, or simply shift their range, forming different aggregation patterns of biodiversity. They can also invoke different strategies of population dynamics to release the pressure, e.g. population fluctuation. Evolutionary pressure, in contrast, can affect the traits of both consumers and resources and consequently often determines the availability and accessibility of resources at a physiological level. Species can modify their functional traits convergently or divergently by changing their morphology, phenology, tolerance, performance and plasticity; such changes are reinforced by heritable genotypes, leading to diverse evolutionary trajectories. Consequently, the architecture of an ecological network is also shaped by the imprint of its evolutionary history. Both evolutionary history and the introduction of alien species bring species with different traits into play and act to craft the architecture of an ecological network, whereas adaptive interaction switching, species rewiring or rapid trait evolution allow the system to be refined through ecological fitting. Network architecture is thus a by-product of both coevolution and ecological fitting, through spontaneous self-organisation, in response to anthropogenic challenges facing embedded socioecological systems (Hui and Richardson 2017).

References

Allesina S, Tang S (2012) Stability criteria for complex ecosystems. Nature 483:205–208

Almeida-Neto M, Ulrich W (2011) A straightforward computational approach for measuring nestedness using quantitative matrices. Environ Model Softw 26:173–178

Almeida-Neto M, Guimarães P, Guimarães PR Jr, Loyola RD, Ulrich W (2008) A consistent metric for nestedness analysis in ecological systems: reconciling concept and measurement. Oikos 117:1227–1239

Atmar W, Patterson BD (1993) The measure of order and disorder in the distribution of species in fragmented habitat. Oecologia 96:373–382

Barabási AL, Albert R (1999) Emergence of scaling in random networks. Science 286:509–512

Bascompte J, Jordano P (2006) The structure of plant-animal mutualistic networks. In: Pascual M, Dunne J (eds) Ecological networks. Linking structure to dynamics in food webs. Oxford University Press, New York, pp 143–159

Bascompte J, Jordano P, Melián CJ, Olesen JM (2003) The nested assembly of plant-animal mutualistic networks. Proc Natl Acad Sci U S A 100:9383–9387

Berlow EL, Neutel AM, Cohen JE, De Ruiter PC, Ebenman BO, Emmerson M, Fox JW, Jansen VAA, Jones JI, Kokkoris GD, Logofet DO, McKane AJ, Montoya JM, Petchey O (2004) Interaction strengths in food webs: issues and opportunities. J Anim Ecol 73:585–598

Camacho J, Guimerà R, Amaral LAN (2002) Robust patterns in food web structure. Phys Rev Lett 88:228102

Cattin MF, Bersier LF, Banašek-Richter C, Baltensperger R, Gabriel JP (2004) Phylogenetic constraints and adaptation explain food-web structure. Nature 427:835–839

Chen X, Cohen JE (2001) Global stability, local stability and permanence in model food webs. J Theor Biol 212:223–235

Cohen JE, Briand F, Newman CM (1990) Community food webs: data and theory. Springer, Berlin

Dieckmann U, Doebeli M (1999) On the origin of species by sympatric speciation. Nature 400:354–357

Doebeli M, Dieckmann U (2000) Evolutionary branching and sympatric speciation caused by different types of ecological interactions. Am Nat 156:S77–S101

Dunne JA, Williams RJ, Martinez ND (2002a) Food-web structure and network theory: the role of connectance and size. Proc Natl Acad Sci U S A 99:12917–12922

Dunne JA, Williams RJ, Martinez ND (2002b) Network structure and biodiversity loss in food webs: robustness increases with connectance. Ecol Lett 8:558–567

Elton CS (1958) Ecology of invasions by animals and plants. Chapman and Hall, London

Erdős P, Rényi A (1959) On random graphs. Publ Math 6:290–297

Fossette S, Glleiss AC, Casey JP, Lewis AR, Hays GC (2012) Does prey size matter? Novel observations of feeding in the leatherback turtle (*Dermochelys coriacea*) allow a test of predator-prey size relationships. Biol Lett 8:351–354

Gell-Mann M (1995) What is complexity? Complexity 1:16–19

Gravel D, Massol F, Leibold M (2016) Stability and complexity in model meta-ecosystems. Nat Commun 7:12457

Guimerà R, Amaral LAN (2005) Cartography of complex networks: modules and universal roles. J Stat Mech Theory Exp 2005:P02001

Hagen M, Kissling WD, Rasmussen C, Carstensen DW, Dupont YL, KaiserBunbury CN, O'Gorman EJ, Olesen JM, MAM DA, Brown LE, AlvesDos-Santos I, Guimarães PR, Maia KP, Marquitti FMD, Vidal MM, Edwards FK, Genini J, Jenkins GB, Trøjelsgaard K, Woodward G, Jordano P, Ledger ME, Mclaughlin T, Morellato LPC, Tylianakis JM (2012) Biodiversity, species interactions and ecological networks in a fragmented world. Adv Ecol Res 46:89–120

Hubbell SP (2001) The unified neutral theory of biodiversity and biogeography. Princeton University Press, Princeton

Hui C, Richardson DM (2017) Invasion dynamics. Oxford University Press, Oxford, UK

Hui C, Richardson DM, Landi P, Minoarivelo HO, Garnas J, Roy HE (2016) Defining invasiveness and invasibility in ecological networks. Biol Invasions 18:971–983

Jordano P, Bascompte J, Olesen JM (2003) Invariant properties in coevolutionary networks of plant animal interactions. Ecol Lett 6:69–81

Kaiser-Bunbury CN, Muff S, Memmott J, Müller CB, Caflisch A (2010) The robustness of pollination networks to the loss of species and interactions: a quantitative approach incorporating pollinator behaviour. Ecol Lett 13:442–452

Kondoh M (2003) Foraging adaptation and the relationship between food-web complexity and stability. Science 299:1388–1391

Krause AE, Frank KA, Mason DM, Ulanowicz RE, Taylor WW (2003) Compartments revealed in food-web structure. Nature 426:282–285

Landi P, Piccardi C (2014) Community analysis in directed networks: in-, out-, and pseudocommunities. Phys Rev E 89:012814

Landi P, Minoarivelo HO, Brännström Å, Hui C, Dieckmann U (2018) Complexity and stability of adaptive ecological networks: a survey of the theory in community ecology. In: Mensah P, Katerere D, Hachigonta S, Roodt A (eds) Systems analysis approach for complex global challenges. Springer, Cham, pp 209–248

Leibold MA, Chase JM (2018) Metacommunity ecology. Princeton University Press, Princeton

MacArthur RH (1955) Fluctuations of animal populations and a measure of community stability. Ecology 36:533–536

May RM (1972) Will a large complex system be stable? Nature 238:413–414

May RM (1973) Stability and complexity in model ecosystems. Princeton University Press, Princeton

McCann KS (2000) The diversity-stability debate. Nature 405:228–233

McCann KS, Hastings A, Huxel GR (1998) Weak trophic interactions and the balance of nature. Nature 395:794–798

Milgram S (1967) The small world problem. Psychology Today 1:61–67

Minoarivelo HO, Hui C (2016a) Trait-mediated interaction leads to structural emergence in mutualistic networks. Evol Ecol 30:105–121

Minoarivelo HO, Hui C (2016b) Invading a mutualistic network: to be or not to be similar. Ecol Evol 6:4981–4996

Minoarivelo HO, Hui C, Terblance JC, Kosakovsky P, Sheffler K (2014) Detecting phylogenetic signal in mutualistic interaction networks using a Markov process model. Oikos 123:1250–1260

Moore JC, Hunt HW (1988) Resource compartmentation and the stability of real ecosystems. Nature 333:261–263

Neutel A-M, Heesterbeek JAP, de Ruiter PC (2002) Stability in real food webs: weak links in long loops. Science 296:1120–1123

Newman M, Girvan M (2004) Finding and evaluating community structure in networks. Phys Rev E 69:026113

Nuwagaba S, Zhang F, Hui C (2015) A hybrid behavioural rule of adaptation and drift explains the emergent architecture of antagonistic networks. Proc R Soc B 282:20150320

Nuwagaba S, Zhang F, Hui C (2017) Robustness of rigid and adaptive networks to species loss. PLoS One 12:e0189086

Okuyama T, Holland JN (2008) Network structural properties mediate the stability of mutualistic communities. Ecol Lett 11:208–216

Olesen JM, Bascompte J, Dupont YL, Jordano P (2007) The modularity of pollination networks. Proc Natl Acad Sci U S A 104:19891–19896

Paine RT (1980) Food webs: linkage, interaction strength and community infrastructure. J Anim Ecol 49:667–685

Pimm SL (1979) Complexity and stability: another look at MacArthur's original hypothesis. Oikos 33:251–257

Rezende EL, Lavabre JE, Guimarães PR, Jordano P, Bascompte J (2007) Non-random coextinctions in phylogenetically structured mutualistic networks. Nature 448:925–928

Rohr RP, Saavedra S, Bascompte J (2014) On the structural stability of mutualistic systems. Science 345:1253497

Rosvall M, Bergstrom CT (2007) An information-theoretic framework for resolving community structure in complex networks. Proc Natl Acad Sci U S A 104(18):7327–7331

Sanchez A (2015) Fidelity and promiscuity in an ant-plant mutualism: a case study of Triplaris and Pseudomyrmex. PLoS One 10:e0143535

Santamaría L, Rodríguez-Gironés MA (2007) Linkage rules for plant-pollinator networks: trait complementarity or exploitation barriers? PLoS Biol 5:354–362

Schelling M, Hui C (2015) modMax: Community structure detection via modularity maximization. R package, version 1.0, cran.r-project.org

Solé RV, Valls J (1992) On structural stability and chaos in biological systems. J Theor Biol 155:87–102

Staniczenko P, Lewis OT, Jones NS, Reed-Tsochas F (2010) Structural dynamics and robustness of food webs. Ecol Lett 13:891–899

Summerhayes VS, Elton CS (1923) Contributions to the ecology of Spitzbergen and Bear Island. J Ecol 11:214–286

Suweis S, Simini F, Banavar JR, Maritan A (2013) Emergence of structural and dynamical properties of ecological mutualistic networks. Nature 500:449–452

Suweis S, Grilli J, Banavar JR, Allesina S, Maritan A (2015) Effect of localization on the stability of mutualistic ecological networks. Nat Commun 6:10179

Valdovinos FS, Ramos-Jiliberto R, Garay-Narváez L, Urbani P, Dunne JA (2010) Consequences of adaptive behaviour for the structure and dynamics of food webs. Ecol Lett 13:1546–1559

van Altena C, Hemerik L, de Ruiter PC (2016) Food web stability and weighted connectance: the complexity stability debate revisited. Theor Ecol 9:49–58

van Baalen M, Křivan V, van Rijn PC, Sabelis MW (2001) Alternative food, switching predators, and the persistence of predator-prey systems. Am Nat 157:512–524

Waser NM (2015) Competition for pollination and the evolution of flowering time. Am Nat 185:iii–iiv

Watts DJ, Strogatz SH (1998) Collective dynamics of small-world networks. Nature 393:440–442

Williams RJ, Martinez ND (2000) Simple rules yield complex food web. Nature 404:180–183

Zhang F, Hui C (2014) Recent experience-driven behaviour optimizes foraging. Anim Behav 88:13–19

Zhang F, Hui C, Terblanche JS (2011) An interaction switch predicts the nested architecture of mutualistic networks. Ecol Lett 14:797–803

Zhang F, Hui C, Pauw A (2013) Adaptive divergence in Darwin's race: how coevolution can generate trait diversity in a pollination system. Evolution 67:548–560

Index